实用
计算机英语
简明教程

丁海燕 编著

清华大学出版社
北京

内容简介

本书按照计算机体系结构组织内容，涉及计算机系统、技术和应用等方面，目的是提高读者计算机英语的实际应用能力。

本书选材与时俱进，内容新颖独特，简明扼要，难度适中，取材广泛。新技术与新产品的介绍与工作、生活息息相关，实用性强。图文混排，易用性强，内容与习题多样化，便于教师与学生开展互动式教学，培养专业英语应用能力。

本书适合作为全国各高校信息类（包括计算机科学与技术、软件工程、计算机网络、信息管理等）的专业英语本科教材，也可作为计算机工程人员的参考用书。

本书封面贴有清华大学出版社防伪标签，无标签者不得销售。

版权所有，侵权必究。举报：010-62782989，beiqinquan@tup.tsinghua.edu.cn。

图书在版编目（CIP）数据

实用计算机英语简明教程/丁海燕编著. —北京：清华大学出版社，2017（2021.8重印）
 ISBN 978-7-302-47484-5

Ⅰ.①实… Ⅱ.①丁… Ⅲ.①电子计算机－英语－高等学校－教材 Ⅳ.①TP3

中国版本图书馆CIP数据核字（2017）第139573号

责任编辑：刘向威　李　晔
封面设计：文　静
责任校对：时翠兰
责任印制：宋　林

出版发行：清华大学出版社
　　　　　网　　址：http://www.tup.com.cn, http://www.wqbook.com
　　　　　地　　址：北京清华大学学研大厦A座　　邮　编：100084
　　　　　社 总 机：010-62770175　　　　　　　　邮　购：010-83470235
　　　　　投稿与读者服务：010-62776969, c-service@tup.tsinghua.edu.cn
　　　　　质量反馈：010-62772015, zhiliang@tup.tsinghua.edu.cn
　　　　　课件下载：http://www.tup.com.cn, 010-83470236

印 装 者：三河市少明印务有限公司
经　　销：全国新华书店
开　　本：185mm×260mm　　印　张：15　　字　数：353千字
版　　次：2017年10月第1版　　　　　　　　　　印　次：2021年8月第5次印刷
印　　数：3801～4300
定　　价：35.00元

产品编号：074991-01

前言

计算机专业与英语密切相关，计算机业界人员必须掌握最新技术，英语水平成为决定工作能力的因素之一。

计算机英语其内容具有很强的专业性，与公共英语相比更注重阅读理解能力，只有看懂计算机软硬件安装使用说明手册、计算机屏幕英语提示信息、编程与程序调试过程中的反馈信息才能解决实际问题。只有看懂计算机新技术、新产品的帮助、培训教程，国外原版计算机教材，以及英文文档和技术资料等，才能及时掌握并研究计算机新技术，提高专业水平。

本教材内容分三个部分。第一部分基础篇，以单元（Unit）为单位，共 10 个单元，包括计算机硬件、软件、操作系统、数据、程序设计、计算机科学、多媒体、网络、因特网及在线服务、万维网。每个单元分为术语、单词、短语、缩略词、练习、泛读课文、术语参考译文。习题形式多样，有选择题、匹配题、填空题、翻译题和口语题等。第二部分应用提高篇，内容取材广泛，均选自 Apple、Adobe、Microsoft 等著名计算机公司及国外计算机技术官方网站。按单元介绍国外计算机学习教程，以及计算机新技术、新软件、新产品、新设备等，如 Dreamweaver CC、C 语言教程、iPhone 手机、Apple 计算机、Windows 10 新特性、Office 2016 新功能等。第三部分附录，提供习题参考答案、专业英语样卷、软件水平考试程序员级专业英语部分试题、构词法和基本句型、学术英语写作常用句子、教材词汇表等。

本书选材与时俱进，内容新颖独特，实用性强，叙述简明扼要，难度适中；内容与习题多样化，教学操作性强，易用性强，适合于高校计算机专业英语本科教学使用。本书可以满足不同院校计算机专业英语本科教学需求，也可作为计算机工程人员的参考用书。

由于作者水平有限，敬请读者批评指正。

编 者
2017 年 3 月

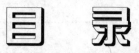

Section I Text

Unit 1 Hardware ·· 3

　Part I　Computer terms ··· 3
　New Words ·· 5
　Phrases ·· 8
　Abbreviation ·· 8
　Exercises ··· 8
　Part II　Reading materials ·· 12
　　　Set up your iPhone, iPad, and iPod touch ···················· 12
　术语参考译文 ·· 14

Unit 2 Software ·· 17

　Part I　Computer terms ··· 17
　New Words ·· 19
　Phrases ·· 20
　Abbreviation ·· 20
　Exercises ··· 20
　Part II　Reading materials ··· 23
　　　Apple Computer ··· 23
　术语参考译文 ·· 24

Unit 3 Operating Systems ·· 26

　Part I　Computer terms ··· 26
　New Words ·· 28
　Phrases ·· 28
　Abbreviation ·· 29
　Exercises ··· 29
　Part II　Reading materials ··· 31
　　　Smartphone ··· 31
　术语参考译文 ·· 32

Unit 4　Data ... 34

Part Ⅰ　Computer terms ... 34
New Words ... 36
Phrases ... 36
Abbreviation ... 37
Exercises ... 37
Part Ⅱ　Reading materials ... 40
　　Data structure ... 40
术语参考译文 ... 41

Unit 5　Programming ... 43

Part Ⅰ　Computer terms ... 43
New Words ... 45
Phrases ... 46
Abbreviation ... 46
Exercises ... 47
Part Ⅱ　Reading materials ... 49
　　The Basics: Object Oriented Programming Concepts ... 49
术语参考译文 ... 50

Unit 6　Computer Science ... 52

Part Ⅰ　Computer terms ... 52
New Words ... 54
Phrases ... 56
Abbreviation ... 56
Exercises ... 57
Part Ⅱ　Reading materials ... 59
　　Artificial Intelligence ... 59
术语参考译文 ... 61

Unit 7　Multimedia ... 64

Part Ⅰ　Computer terms ... 64
New Words ... 66
Phrases ... 67
Abbreviation ... 68
Exercises ... 68
Part Ⅱ　Reading materials ... 71
　　Big data ... 71

术语参考译文 ·· 72

Unit 8　Networks ·· 74

　　Part I　Computer terms ··· 74
　　New Words ·· 76
　　Phrases ··· 77
　　Abbreviation ·· 77
　　Exercises ·· 78
　　Part II　Reading materials ·· 80
　　　　Network security ··· 80
　　术语参考译文 ·· 81

Unit 9　Internet and Online Services ··· 84

　　Part I　Computer terms ··· 84
　　New Words ·· 86
　　Phrases ··· 86
　　Abbreviation ·· 87
　　Exercises ·· 87
　　Part II　Reading materials ·· 90
　　　　Internet ·· 90
　　　　eCommerce Resources ·· 91
　　术语参考译文 ·· 92

Unit 10　World Wide Web ··· 94

　　Part I　Computer terms ··· 94
　　New Words ·· 96
　　Phrases ··· 97
　　Abbreviation ·· 97
　　Exercises ·· 97
　　Part II　Reading materials ·· 100
　　　　Understand Web applications ··· 100
　　术语参考译文 ·· 107

Section II　Computer Courses

Unit 1　Adobe Dreamweaver CC ··· 111

Unit 2　C Programming ·· 119

Unit 3　Features Available only on Windows 10 ·· 139

Unit 4　Excel Training ··· 144

Unit 5　Word Training ··· 153

Unit 6　PowerPoint Training ·· 157

Unit 7　Access Training ·· 161

附　录

附录一　习题参考答案 ··· 171
    Unit 1　Hardware ··· 171
    Unit 2　Software ·· 171
    Unit 3　Operating Systems ··· 173
    Unit 4　Data ··· 173
    Unit 5　Programming ·· 174
    Unit 6　Computer Science ·· 175
    Unit 7　Multimedia ··· 175
    Unit 8　Networks ··· 176
    Unit 9　Internet and Online Services ··· 176
    Unit 10　World Wide Web ··· 177

附录二　专业英语样卷 ··· 178
    试卷一 ··· 178
    试卷一参考答案 ··· 180
    试卷二 ··· 180
    试卷二参考答案 ··· 182
    试卷三 ··· 183
    试卷三参考答案 ··· 187
    试卷四 ··· 187
    试卷四参考答案 ··· 191

附录三　软件水平考试程序员级专业英语试题节选 ··························· 192

附录四　英语构词法和基本句型 ·· 198

附录五　学术英语写作常用句子 ·· 210

附录六　词汇表···212
　单词表···212
　短语表···221
　缩写表···225

Section I Text

Unit 1　　　　　　　　　　Hardware

Part I Computer terms

- **Hardware**

Computer hardware refers to the physical parts that make up the entirety of a computer. Hardware includes the electrical, mechanical, data storage and magnetic components among other parts.

- **CPU—central processing unit**

CPU is the abbreviation for central processing unit (the processor). The CPU is the brains of the computer where most calculations take place.

- **Microprocessors**

A microprocessor is a silicon chip that contains the central processing unit. Different types of microprocessors include Motorola microprocessors and Pentium microprocessors.

- **Register**

A special, high-speed storage area within the CPU. All data must be represented in a register before it can be processed.

- **Buses**

Refers to a set of wires through which data is transmitted from one part of the computer to another. There are multiple different types of buses including memory buses, control buses, and internal buses.

- **Motherboards**

The motherboard is the main circuit board of a microcomputer and contains the connectors for attaching additional boards.

- **Connectors, plugs and sockets**

Connector refers to the parts that plug into and out of devices. They are used to connect monitors to laptops, mice to keyboards, as well as external storage devices to computers.

- **Port**

(1) An interface on a computer to which you can connect a device. Personal computers have various types of ports. Internally, there are several ports for connecting disk drives, display screens, and keyboards. Externally, personal computers have ports for connecting modems, printers, mice, and other peripheral devices.

(2) In TCP/IP and UDP networks, an endpoint to a logical connection. The port number

identifies what type of port it is. For example, port 80 is used for HTTP traffic.

- **NIC—network interface card**

Often abbreviated as *NIC*, an *expansion board* you insert into a computer so the computer can be connected to a network. Most NICs are designed for a particular type of network, protocol, and media, although some can serve multiple networks.

- **Data storage**

Refers to information being held in specific areas. Elements of data storage include intelligent information management and advanced technology attachment.

- **Monitor**

Another term for display screen. The term monitor, however, usually refers to the entire box, whereas display screen can mean just the screen. In addition, the term monitor often implies graphics capabilities.

There are many ways to classify monitors. The most basic is in terms of color capabilities, which separates monitors into three classes:

monochrome: Monochrome monitors actually display two colors, one for the background and one for the foreground. The colors can be black and white, green and black, or amber and black.

gray-scale: A gray-scale monitor is a special type of monochrome monitor capable of displaying different shades of gray.

color: Color monitors can display anywhere from 16 to over 1 million different colors. Color monitors are sometimes called RGB monitors because they accept three separate signals—red, green, and blue.

- **Hardware companies**

Hardware companies are businesses that design, produce and sell computer hardware. This includes disk drives, screens, keyboards, mice and even storage arrays and other enterprise systems.

- **Integrated Circuits (ICs)**

Integrated Circuits are chips made from semiconductor material. Integrated circuits are associated with multi-port memories, nanotechnology, as well as silicon.

- **Peripheral device**

Peripherals are computer devices, such as a CD-ROM drive or printer, that is not part of the essential computer.

- **Input device**

Refers to any type of machine that feeds data into a computer. Examples of input devices include keyboards, monitors, and mice.

- **Output device**

Any machine capable of representing information from a computer. This includes display screens, printers, plotters, and synthesizers.

- **Interface**

A boundary across which two independent systems meet and act on or communicate with each other. In computer technology, there are several types of interfaces.

- **Memory**

The term memory identifies data storage that comes in the form of chips, and the word storage is used for memory that exists on tapes or disks. Moreover, the term memory is usually used as a shorthand for physical memory, which refers to the actual chips capable of holding data. Some computers also use virtual memory, which expands physical memory onto a hard disk.

Every computer comes with a certain amount of physical memory, usually referred to as *main memory* or *RAM*. You can think of main memory as an array of boxes, each of which can hold a single byte of information. There are several different types of memory: RAM, ROM, and flash.

- **RAM—random access memory**

RAM is an acronym for random access memory. It is the most common type of memory found in computers and other devices, such as printers.

- **ROM—read-only memory**

Acronym for read-only memory. ROM is computer memory on which data has been prerecorded. It retains its contents even when the computer is turned off.

- **Memory address**

A number that is assigned to each byte in a computer's memory that the CPU uses to trace where data and instructions are stored in RAM. Each byte is assigned a memory address whether or not it is being used to store data.

- **Keyboard**

A keyboard enables you to enter data into a computer and other devices. Typical keys are classified as alphanumeric, punctuation and special keys.

- **Mouse**

A device that controls the movement of the cursor or pointer on a display screen. A mouse is a small object you can roll along a hard, flat surface.

- **Printer**

A printer is any device that prints text or illustrations on paper. There are many different types of printers: laser printer, impact printer, ink-jet printer and thermal printer and so on.

New Words

abbreviation	[əˌbriːviˈeiʃn]	*n.* 省略，缩写，简化，缩写词
abbreviate	[əˈbriːvieit]	*vt.* 缩略；使简短；缩简
electrical	[iˈlektrikl]	*adj.* 用电的，与电有关的，电学的

mechanical	[mi'kænikəl]	adj. 机械的，机械学的
component	[kəm'pəunənt]	n. 成分；零件；[数]要素 adj. 组成的；构成的
calculation	[ˌkælkjə'leʃn]	n. 计算；盘算；估计
silicon	['silikən]	n. [化]硅；硅元素
chip	[tʃip]	n. 芯片
represent	[ˌrepri'zent]	vt. 表现，象征；代表，代理
motherboard	['mʌðəbɔ:d]	n. 主板，母板
register	['redʒistə]	n. 记录；登记簿；登记 vt. & vi. 登记；注册
bus	[bʌs]	n. [计]总线；公共汽车
microprocessor	[ˌmaikrou'prousesə]	n. 微处理器
connector	[kə'nektə(r)]	n. 连接器，连接体
plug	[plʌg]	n. 插头；塞子 vt. & vi. 插入；塞住
socket	['sɔkit]	n. 插座；灯座；窝，穴
input	['input]	n. & vt. 输入
output	['autput]	n. & vt. 输出
port	[pɔ:t]	n. 港口；（计算机与其他设备的）接口
disk drive	[disk draiv]	n. 磁盘驱动器
peripheral	[pə'rifərəl]	adj. 外围的
screen	[skri:n]	n. 屏幕；银幕
modem	['məudem]	n. 调制解调器
identify	[ai'dentifai]	vt. 确定；识别；认出 vi. 确定；认同
expansion	[ik'spænʃn]	n. 扩张；扩大；扩展
board	[bɔ:d]	n. 板；董事会 vt. 上（船、车或飞机）
protocol	['proutəkɔ:l]	n. 礼仪；（数据传递的）协议
monitor	['mɔnitə(r)]	n. 显示屏 vt. 监督；监控 vi. 监视
separate	['sepəreit]	vt. & vi. 分开；（使）分离；隔开
monochrome	['mɔnəkrəum]	n. 单色画，黑白照片 adj. 单色的，黑白的
gray-scale	[grei skeil]	n. 灰度
color	['kʌlə]	n. 颜色，色彩
background	['bækgraund]	n. 背景；底色；背景资料；配乐

foreground	[ˈfɔːgraund]	n. 前景
integrate	[ˈintigreit]	vt. 使一体化；使整合 vi. 成为一体
circuit	[ˈsəːkit]	n. 电路
semiconductor	[ˌsemikənˈdʌktə(r)]	n. 半导体
material	[məˈtiəriəl]	n. 素材；材料，原料
associate	[əˈsəuʃieit]	vt. 联想；(使)发生联系；(使)联合；结交
nanotechnology	[ˌnænəutekˈnɔlədʒi]	n. 纳米技术，毫微技术
mice	[mais]	n. 老鼠(的名词复数)；鼠标
plotter	[ˈplɔtə(r)]	n. 绘图仪
synthesizer	[ˈsinθiˌsaizə]	n. 合成物；合成器
interface	[ˈintəfeis]	n. 界面；[计]接口
memory	[ˈmeməri]	n. 记忆；[计]存储器；内存
address	[əˈdres]	n. 地址；称呼；演说；通信处
expand	[ikˈspænd]	vt. 扩张；使……变大 vi. 扩展
shorthand	[ʃɔːthænd]	n. 速记
virtual	[ˈvɜːtʃuəl]	adj. 实质上的；虚拟的
memory	[ˈmeməri]	n. 存储器
flash	[flæʃ]	n. 闪光 vt.& vi. 使闪光，使闪烁
trace	[treis]	vt. 跟踪，追踪；追溯
assign	[əˈsain]	vt. 分派，选派，分配；归于
instruction	[inˈstrʌkʃən]	n. 授课；指令
alphanumeric	[ˌælfənjuːˈmerik]	adj. 文字数字的
punctuation	[ˌpʌŋktʃuˈeʃən]	n. 标点符号
cursor	[ˈkɜːsə(r)]	n. 光标
pointer	[ˈpɔintə]	n. 线索；指针
flat	[flæt]	n. 平面；公寓 adj. 平的；单调的
keyboard	[ˈkiːbɔːd]	n. 键盘；琴键
mouse	[maus]	n. 鼠标；老鼠
laser	[ˈleizə]	n. 激光
impact	[ˈimpækt]	n. 影响；冲击 vt. 撞击；压紧

Hardware

| jet | [dʒet] | n. 喷嘴，喷雾 vt. 喷射，喷出 |
| thermal | [ˈθəːməl] | adj. 热的，温热的 |

Phrases

memory buses	内存总线
control buses	控制总线
internal buses	内部总线
external storage device	外存设备
integrated circuit	集成电路
peripheral device	外围设备
input device	输入设备
output device	输出设备
virtual memory	虚拟内存
main memory	主存
memory address	内存地址
display screen	显示屏
flat surface	平面
laser printer	激光打印机
impact printer	击打式打印机
ink-jet printer	喷墨打印机
thermal printer	热敏打印机

Abbreviation

IC—Integrated Circuits　　　　　　　　集成电路
CPU—Central Processing Unit　　　　　中央处理器
NIC—Network Interface Card　　　　　网络接口卡
RAM—Random Access Memory　　　　随机存储器
ROM—Read-Only Memory　　　　　　只读存储器
CD-ROM—Compact Disk- Read-Only Memory　只读碟

Exercises

Ex1. Multiple choice.

1. A (　　) computer is a personal computer whose hardware is capable of using any or all

of the following media in a program: audio, text, graphics, video and animation.

    A. database   B. multimedia   C. network   D. mainframes

2. The ( ) controls the cursor or pointer on the screen and allows the user to access commands by pointing and clicking.

    A. graphics   B. printer   C. program   D. mouse

3. A ( ) copies a photograph, drawing or page of text into the computer.

    A. scanner   B. printer   C. display   D. keyboard

4. Files can be lost or destroyed accidentally. Keep ( ) copies of all data on removable storage media.

    A. backup   B. back   C. black   D. backdown

5. A ( ) is a functional unit that interprets and carries out instructions.

    A. memory   B. processor   C. storage   D. network

6. Insufficient ( ) can cause a processor to work at 50% or even more below its performance potential.

    A. mouse   B. I/O   C. document   D. memory

7. The ( ) is a temporary storage area that you can use to copy or move selected text or object among applications.

    A. cache   B. pool   C. buffer   D. clipboard

8. A typical peripheral device has ( ) which the processor uses to select the device's internal registers.

    A. data   B. a control   C. a signal   D. an address

9. ( ) is a device that converts images to digital format.

    A. Copier   B. Printer   C. Scanner   D. Display

10. A computer is a programmable machine with two principle characteristics. Firstly, it responds to a specific set of __(1)__ in a well-defined manner. Secondly, it can execute a prerecorded list of instructions, which is referred to as a program. Modern computers are electronic and digital. The actual machinery, including wires, transistors, and integrated circuits, is called __(2)__; and the instructions and data are called __(3)__.

All general-purpose computers require the following hardware components:

__(4)__, which enables a computer to store, at least temporarily, data and program.

__(5)__, which allows a computer to permanently retain large amounts of data, including disk drives, tape drives, and laser discs.

__(6)__, which, in the form of a keyboard and a mouse, is the conduit (cannel) through which data and instruction enter a computer.

__(7)__, which covers such equipment as the display screen, the printer, and various other devices through which you can get what the computer has accomplished.

__(8)__, which stands as the heart, or you may just as well as say, the brain of the computer, carrying out instructions provided either by the user or the system itself.

In addition to these components, many others make it possible for the basic components to work together efficiently. For example, every computer requires a bus that transmits data from one part of the computer to another. Common bus types include ISA, (9), VESA, and a few others.

Words to be chosen from:
A. instructions   B. CPU   C. software   D. hardware   E. memory
F. mass storage device   G. input device   H. output device   I. PCI

Answers: (1) _____ (2) _____ (3) _____ (4) _____ (5) _____ (6) _____
(7) _____ (8) _____ (9) _____

11. Memory is the electronic holding place for (1) and data that your computer's microprocessor can reach quickly. When your computer is in (2) operation, its memory usually contains the main parts of the OS and some or all of the application programs and related data that are being used. Memory is often used as a shorter acronym for random access memory(RAM). This kind of memory is (3) on one or more microchips that are physically close to the microprocessor in your computer. Most desktop and notebook computers sold today include at least 32 megabytes of RAM, and are (4) to more. The more RAM you have, the (5) frequently the computer has to access instructions and data from the more slowly accessed hard disk.

Words to be chosen from:
A. less   B. upgradable   C. located   D. instruction   E. normal

Answers: (1) _____ (2) _____ (3) _____ (4) _____ (5) _____

Ex2. Match each numbered item with the most closely related lettered item.

a. monitor   b. secondary storage device   c. CPU   d. hard disks
e. microcomputers   f. optical disks   g. output device
h. input device   i. computer network   j. primary storage

(   )1. The electronic circuitry within a computer that carries out the instructions of a computer program by performing the basic arithmetic, logical, control and input/output(I/O) operations specified by the instructions.

(   ) 2. A screen that displays peripheral output to the user.

(   ) 3. The least powerful and most widely used type of computer.

(   ) 4. Communications system connecting two or more computers .

(   ) 5. Translates data and programs that humans can understand into a form that the computer can process.

(   ) 6. Translates processed information from the computer into a form that humans can understand.

(   ) 7. Holds data and program instructions for processing data.

(   ) 8. Holds data and programs even after electrical power to the system has been turned off.

(   ) 9. Typically used to store programs and very large data files.

(   ) 10. Use laser technology and have the greatest capacity of all secondary storage.

Ex3. Computer English test.

1. A computer system consists of several basic __(1)__. An input device provides data. The data are stored in __(2)__, which also holds a program. Under control of that program the computer's __(3)__ manipulates the data, storing the results back into memory. Finally, the results flow from the computer to an __(4)__ device. Additionally, most modern computers use secondary storage to extend memory __(5)__.

    (1) A. parts   B. components   C. ingredient   D. assembly
    (2) A. floppy disk   B. hard disk   C. memory   D. tape
    (3) A. processor   B. heart   C. controller   D. plate
    (4) A. input   B. output   C. import   D. export
    (5) A. capability   B. capacity   C. content   D. size
Answers: (1) _____ (2) _____ (3) _____ (4) _____ (5) _____

2. The processor, often called the central processing unit (CPU) or main processor, is the component that processes or manipulates __(1)__. A processor can do nothing __(2)__ a program to provide control; whatever intelligence a computer has is derived from __(3)__ not __(4)__. The processor manipulates data stored in main memory __(5)__ the control of a program store in main memory.

    (1) A. data   B. program   C. figure   D. number
    (2) A. with   B. without   C. or   D. but
    (3) A. program   B. hardware   C. software   D. processor
    (4) A. processor   B. software   C. memory   D. hardware
    (5) A. at   B. under   C. in   D. on
Answers: (1) _____ (2) _____ (3) _____ (4) _____ (5) _____

3. Secondary storage is an __(1)__ of main memory, not a __(2)__ for it. A computer cannot execute a program stored on disk __(3)__ it is first copied into main memory. A computer cannot manipulate the data stored on a secondary medium until they have been copied into main memory. Main memory holds the current program and the current data; secondary storage is __(4)__ storage.

    (1) A. extension   B. extend   C. accessary   D. intension
    (2) A. tool   B. replacement   C. branch   D. backup
    (3) A. in case   B. if   C. lest   D. unless
    (4) A. temporary   B. shortterm   C. longterm   D. eternal
Answers: (1) _____ (2) _____ (3) _____ (4) _____

Ex4. Translate the following sentences into English.

1. 处理数据以产生信息的设备称为硬件，包括键盘、鼠标、显示器、主机和其他设备。硬件由软件控制。

2. 计算机是能够遵循指令以接收输入、处理输入并产生信息的电子设备。

3. 微机是功能最弱但使用最广且发展最快的计算机类型，微机的类型包括台式机、笔记本电脑和 PDA。

4. 输入设备将人能理解的数据和程序翻译成计算机能处理的格式。

5. 辅存设备不像内存,即使计算机系统的电源被关闭也能保存数据和程序。

Ex5. Speaking.

1. Explain the five parts of computer hardware and their main functions.

2. What is the difference between input and output? What are the most common input devices? What are the most common output devices?

3. What are the main parts of the CPU?

4. What is RAM?

5. What is the typical unit used to measure RAM memory and storage memory?

6. What is ALU? What does it do?

7. How can we store data and programs permanently?

Part II   Reading materials

Set up your iPhone, iPad, and iPod touch

- **Turn on your device**

You'll see "Hello" in many languages. Press the Home button to unlock your device and begin set up.

If you're blind or have low vision, you can turn on VoiceOver or Zoom from the Hello screen.

Get help if your device won't turn on, or if it's disabled or requires a passcode.

- **Select your language and country**

When asked, choose your language. Then tap your country or region. This affects how information looks on your device, including date, time, contacts, and more.

- **Activate your device**

You need to connect to a Wi-Fi network, cellular network, or iTunes to activate and continue setting up your device.

Tap the Wi-Fi network that you want to use or select a different option. If you're setting up an iPhone or iPad (Wi-Fi + Celluar), you might need to insert your SIM card first.

Then decide on Location Services, a feature you need for apps like Maps and Find My Friends.

Get help if you can't connect to Wi-Fi or if you can't activate your iPhone.

- **Set up Touch ID and create a passcode**

On some devices, you can set up Touch ID (ref. Fig.1-1-1). With this feature, you can use your fingerprint to unlock your device and make purchases.

Next, set a six-digit passcode to help protect your data. You need a passcode to use features like Touch ID and Apple Pay. If you'd like a four-digit passcode, custom passcode, or no passcode, tap Passcode Options.

- **Restore or transfer your information and data**

If you have an iCloud or iTunes backup, or an Android device, you can restore or transfer your data from your old device to your new device.

If you don't have a backup or another device, select Set Up as New iPhone.

- **Sign in with your Apple ID and set up iCloud Drive**

Enter your Apple ID (ref. Fig.1-1-2) and password, or tap Don't have an Apple ID or forgot it. From here, you can recover your Apple ID or password, create an Apple ID, or set it up later.

If you use more than one Apple ID, tap Use different Apple IDs for iCloud and iTunes.

Then accept the iOS Terms and Conditions. If you signed in with your Apple ID, follow the steps to set up iCloud Drive, Apple Pay, and iCloud Keychain.

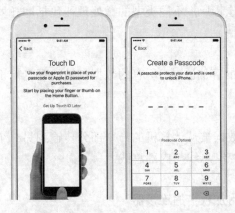

Fig.1-1-1   Touch ID

Fig.1-1-2   Apple ID

- **Set up Siri**

Choose whether to use Siri. On some devices, you'll be asked to speak some phrases so that Siri can get to know your voice.

- **Customize the click of your Home button**

If you have an iPhone 7 or iPhone 7 Plus, you can set up how your Home button responds when you press it. Just tap an option, then press the Home button to give it a try. To set the one you like best, tap Next.

If you want to skip this screen or want to set up your Home button later, tap Customize Later in Settings.

- **Choose settings for app analytics and display resolution**

Decide whether to share information with app developers.

Next, if you have an iPhone 6 or later, choose between two display resolutions: Standard shows more on your screen, while Zoomed uses larger text and controls. Tap Next to continue.

- **Finish up**

Tap Get Started to begin using your device. Make a safe copy of your data by backing up, and learn about more features in the user guide for your iPhone, iPad, or iPod touch.

术语参考译文

硬件
计算机硬件是指构成计算机整体的物理部件。硬件包括各部件之间的电气、机械、数据存储和磁性元件。

CPU——中央处理器
CPU 是中央处理器（处理器）的缩写。CPU 是计算机的大脑，大多数计算都发生在 CPU 当中。

微处理器
微处理器是一个包含中央处理器的硅芯片，不同类型的微处理器包括摩托罗拉微处理器和奔腾微处理器。

寄存器
寄存器是 CPU 中一个特殊的、高速的存储区域。所有的数据在被处理之前都要在寄存器中表示。

总线
总线是指一组线路，通过这组线路数据被从计算机的一个部分传输到另一个部分。有多种不同类型的总线，包括内存总线、控制总线与内部总线。

母板
母板是微机的主电路板，包含连接附加板卡的连接器。

连接器、插头与插座
连接器是指插入与拔出设备的部件。连接器被用于将显示器与笔记本电脑相连，将鼠标与键盘相连，以及将外部存储设备与计算机相连。

端口
（1）计算机上的一个接口，通过该接口可以连接一个设备。个人计算机有多种类型的端口。在内部，有几个端口，用于连接磁盘驱动器、显示器和键盘。在外部，个人计算机有连接调制解调器、打印机、鼠标和其他外围设备的端口。

（2）在 TCP/IP 与 UDP 网络中，端口是指逻辑连接的端点。端口号标识是什么类型的端口。例如，端口 80 用于 HTTP 传输。

NIC——网络接口卡
网络接口卡常常缩写为 NIC（一般简称为网卡），是插入计算机的一个扩展卡，使得计算机能够与网络连接。大多数网卡是为某一特定类型的网络、协议和传输介质设计的，尽管有一些网卡可以同时用于多个网络。

数据存储
数据存储是指信息被保存在特定的区域。数据存储的元素包括智能信息管理与先进的技术附件。

显示器
显示屏的另一个术语，不过显示器这个术语通常是指整个箱子，而显示屏仅是指屏幕。另外，术语显示器通常意味着图形能力。

有多种显示器分类的方法。最基本的是所谓的颜色能力，它将显示器分成三类。
- 单色：单色显示器实际上显示两种颜色，其中一种用于背景色，另一种用于前景色。颜色可以是黑色与白色、绿色与黑色或者琥珀与黑色。
- 灰度：一个灰度显示器是一种特殊类型的单色显示器，能够显示不同深浅的灰色。
- 彩色：彩色显示器可以显示任何从 16 种到超过 100 万种不同的颜色。彩色显示器有时被称为 RGB 显示器，因为它们接收三个独立的信号——红、绿、蓝。

硬件公司
硬件公司是设计、生产和销售计算机硬件的企业。这包括磁盘驱动器、屏幕、键盘、鼠标，甚至存储阵列和其他企业系统。

集成电路
集成电路是由半导体材料制成的芯片。集成电路与多端口存储器、纳米技术以及硅有关。

外围设备
外围设备是计算机设备，如 CD-ROM 驱动器或打印机，它们不是基本计算机的一部分。

输入设备
输入设备是指将数据输入计算机的任何类型的机器。输入设备的例子包括键盘、监视器和鼠标。

输出设备
任何能够表示计算机中信息的机器，包括显示器、打印机、绘图仪和合成器。

接口
两个独立系统相遇、相互作用或相互通信的边界。在计算机技术中，有几种类型的接口。

内存
术语内存标识芯片形式的数据存储器，而单词存储器用于磁带或磁盘存储器。此外，术语内存通常用作物理内存的简写，它指的是能够存储数据的实际芯片。一些计算机也使用虚拟内存，它将物理内存扩展到硬盘上。

每台计算机都有一定的物理内存，通常称为主存或 RAM。你可以把主存看作是一组盒子，每个盒子都可以容纳一个字节的信息。有几种不同类型的内存：RAM、ROM 和闪存。

随机存储器
RAM 是随机存储器的首字母缩写。它是计算机和其他设备中最常见的存储器类型，例如打印机。

ROM——只读存储器
只读存储器的首字母缩写。ROM 是其数据预先已经记录在内的计算机内存。即使计算机关闭，它仍保留其内容。

内存地址
一个数字，被分配给计算机内存中的每个字节，CPU 用内存地址来跟踪 RAM 中的数据和指令的位置。每个字节被分配一个内存地址，无论它是否被用来存储数据。

键盘

键盘用于将数据输入计算机和其他设备。典型的按键分为字母数字、标点符号和特殊键。

鼠标

控制屏幕上光标或指针移动的装置。鼠标是一个小物体，你可以在一个坚硬的平面上滚动它。

打印机

打印机是在纸上打印文本或插图的任何设备。有许多不同类型的打印机：激光打印机、击打式打印机、喷墨打印机和热敏打印机等。

Unit 2　　Software

Part I　Computer terms

- **Application (application software)**

A software application is a program or group of programs designed for end users. Applications can be systems software or applications software.

- **GUI—graphical user interface**

Graphical user interface (GUI) takes advantage of the computer's graphics capabilities to make the program easier to use with pointing devices, menus and icons.

- **Desktop**

In graphical user interfaces, a *desktop* is the metaphor used to portray file systems. Such a desktop consists of pictures, called *icons*, that show files, folders, and various types of documents (that is, letters, reports, pictures).

- **Integrated development environment**

Abbreviated as *IDE*, a programming environment integrated into a software application that provides a GUI builder, a text or code editor, a compiler and/or interpreter and a debugger. Visual Studio, Delphi, JBuilder, FrontPage and DreamWeaver are all examples of IDEs.

- **Menu**

A list of commands or options from which you can choose.

- **Dialog box**

A box that appears on a display screen to present information or request input.

- **Contextual menu**

When using an application or an operating system, the menu that appears when you click on the right-hand button of a two-button mouse (also called *right clicking*). Single-button mouse users can bring up the contextual menu by holding down the *Ctrl* key while clicking.

- **Alignment**

(1) In text, the arrangement of text or graphics relative to a margin.

(2) In reference to graphical objects, alignment describes their relative positions.

- **Boldface**

A font that is darker than the regular face. Most word processors allow you to mark text as boldface.

- **Font**

A design for a set of characters. A font is the combination of typeface and other qualities, such as size and spacing.

- **Margins**

In word processing, the strips of white space around the edge of the paper. Most word processors allow you to specify the widths of margins.

- **Word wrap**

In word processing, a feature that causes the word processor to force all text to fit within the defined margins.

- **Indent**

The word indent is used to describe the distance, or number of blank spaces used to separate a paragraph from the left or right margins.

- **Default**

A value or setting that a device or program automatically selects if you do not specify a substitute. For example, word processors have default margins and default page lengths that you can override or reset.

- **Style sheet**

In word processing and desktop publishing, a style sheet is a file or form that defines the layout of a document. When you fill in a style sheet, you specify such parameters as the page size, margins, and fonts.

- **Left justify**

To align text along the left margin. Left-justified text is the same as flush-left text.

- **Image dimensions**

Image dimensions are the length and width of a digital image. It is usually measured in pixels, but some graphics programs allow you to view and work with your image in the equivalent inches or centimeters.

- **Marquee**

On web pages, a scrolling area of text. Starting with Version 2, Microsoft Internet Explorer supports a special <MARQUEE> tag for creating these areas.

- **Filter**

A program that accepts a certain type of data as input, transforms it in some manner, and then outputs the transformed data.

- **Color saturation**

In graphics and imaging, color saturation is used to describe the intensity of color in the image. A saturated image has overly bright colors.

New Words

capability	[ˌkeipəˈbiləti]	n. 性能；容量；才能；能力
pointing device	[ˈpɔintiŋ diˈvais]	n. 点击设备
icon	[ˈaikɔn]	n. 图标；图符
desktop	[ˈdesktɔp]	n. 桌面
metaphor	[ˈmetəfə(r)]	n. 象征；隐喻；暗喻
portray	[pɔːˈtrei]	n. 描绘；描述；画像
folder	[ˈfəuldə(r)]	n. 文件夹
compiler	[kəmˈpailə(r)]	n. 汇编者；编译程序
interpreter	[inˈtɜːpritə(r)]	n. 解释者；口译译员；解释程序
debugger	[ˌdiːˈbʌgə(r)]	n. 调试器
option	[ˈɔpʃn]	n. 选项；选择权
dialog box	[ˈdaiəlɔg bɔks]	n. 对话框
contextual	[kənˈtekstʃuəl]	adj. 上下文的，前后关系的
alignment	[əˈlainmənt]	n. 队列；排成直线；对齐
margin	[ˈmɑːdʒin]	n. 边缘，页边距
boldface	[ˈbəuldfeis]	n. 黑体字，粗体
font	[fɔnt]	n. 体；字形
typeface	[ˈtaipfeis]	n. 字体；字型
strip	[strip]	n. 长条；条板；带状地带
word wrap	[wəːd ræp]	n. 自动换行
indent	[inˈdent]	n. （印刷中的）缩进；订单
blank space	[blæŋk speis]	n. 空白（区）
default	[diˈfɔːlt]	n. 默认值，缺省 adj. 默认的
substitute	[ˈsʌbstitjuːt]	n. 用……替代 vi. 替代物；代替者
override	[ˌəuvəˈraid]	vt. 覆盖；推翻
style sheet	[stail ʃiːt]	n. 样式表
layout	[ˈleiaut]	n. 布局；安排；设计
parameter	[pəˈræmitə(r)]	n. 参数，决定因素
left justify	[left ˈdʒʌstifai]	n. 左对齐
image	[ˈimidʒ]	n. 图像；肖像
dimension	[diˈmenʃən, dai-]	n. 尺寸；[数]次元，度，维
pixel	[ˈpiksl]	n. 像素
equivalent	[iˈkwivələnt]	adj. 相等的；等价的 n. 对等物
inch	[intʃ]	n. 英寸

centimeter	[ˈsentiˌmiːtə]	n. 厘米
marquee	[mɑːˈkiː]	n. 选取框；大帐篷；跑马灯
scroll	[skrəul]	vt. 滚动；卷页
tag	[tæg]	n. 标签
filter	[ˈfiltə(r)]	n. 滤波器；滤光器；滤镜；过滤器 vt. 过滤
transform	[trænsˈfɔːm]	vt. 变换；改变 vi. 改变
color saturation	[ˈkʌlə ˌsætʃəˈreiʃən]	n. 色彩饱和度
intensity	[inˈtensəti]	n. 强度；烈度

Phrases

contextual menu	上下文菜单，快捷菜单
systems software	系统软件
applications software	应用软件
pointing device	点击设备
code editor	代码编辑程序
graphical object	图形对象
word wrap	自动换行
blank space	空白
default margin	默认边距
page length	页长
word processor	文字处理软件
style sheet	样式表
left justify	左对齐
flush-left	左对齐
image dimension	图像尺寸
color saturation	色彩饱和度

Abbreviation

GUI—Graphical user interface　　　　图形用户界面
IDE—integrated development environment　　集成开发环境

Exercises

Ex1. Multiple choice.

1. (　　) software, also called end-user program, includes database programs, word processors, preadsheets, etc.

    A. Application   B. System   C. Compiler   D. Utility
2. (   ): An error can be caused by attempting to divided by 0.
    A. Interrupt   B. Default   C. Underflow   D. Overflow
3. Integration (   ) is the process of verifying that the components of a system work together as described in the program design and system design specification.
    A. trying   B. checking   C. testing   D. coding
4. (   ) is a contiguous, numbered set of variables of a given base type, which can be used and passed to functions as a unit.
    A. Record   B. Array   C. File   D. Parameter
5. When the result of an operation becomes larger than the limits of the representation, (   ) occurs.
    A. overdose   B. overflow   C. overdraft   D. overexposure
6. (   ) are prewritten formulas that perform calculations automatically.
    A. Functions   B. Macros   C. Templates   D. Calculators
7. In a relational database, data is organized into (   ).
    A. fields   B. columns   C. records   D. tables
    E. rows
8. (   ) are specialized programs designed to allow input and output devices to communicate with a computer system.
    A. Utilities   B. Resources   C. Device drivers D. GUIs
    E. Windows
9. Language translators convert human language into (   ).
    A. machine language   B. UNIX
    C. service programs   D. operating system
10. To remove unneeded programs and related files from a hard disk you would use a (   ).
    A. backup program   B. file compression program
    C. antivirus program   D. uninstall program
11. (   ): A graphical bar with buttons that perform some of the most common commands.
    A. Title bar   B. Tool bar   C. Status bar   D. Scroll bar

Ex2. Match each numbered item with the most closely related lettered item.
    a. window   b. GUI   c. word processor   d. style sheet
    e. applications software   f. find   g. sorting   h. formulas
    i. toolbar   j. spelling checker

(   ) 1. A type of user interface that allows users to interact with electronic devices through graphical icons and visual indicators such as secondary notation, instead of text-based user interfaces, typed command labels or text navigation.

(   ) 2. Rectangular area that can contain a document, program, or message.

(   ) 3. A feature that contains buttons and menus to provide access to commonly used commands.

(　　) 4. A feature in desktop publishing programs that store and apply formatting to text. It is a form of separation of presentation and content.

(　　) 5. Software that creates text-based documents such as reports, letters, and memos.

(　　) 6. Identifies incorrectly spelled words and suggests alternatives.

(　　) 7. Tool that quickly locates any character, word, or phrase in a document.

(　　) 8. Instructions for calculations.

(　　) 9. A computer program designed to perform a group of coordinated functions, tasks, or activities for the benefit of the user, such as a word processor, a spreadsheet, an accounting application, a web browser, or a media player.

(　　) 10. Arranging objects numerically or alphabetically.

Ex3. Computer English test.

1. Software design is a __(1)__ process. It requires a certain __(2)__ of flair on the part of the designer. Design cannot be learned from a book. It must be practiced and learnt by experience and study of existing systems. A well __(3)__ software system is straight-forward to implement and maintain, easily __(4)__ and reliable. Badly __(3)__ software system, although they may work, are __(5)__ to be expensive to maintain, difficult to text and unreliable.

(1) A. create   B. created   C. creating   D. creative
(2) A. amount   B. amounted   C. mount   D. mounted
(3) A. design   B. designed   C. designing   D. designs
(4) A. understand   B. understands   C. understanding   D. understood
(5) A. like   B. likely   C. unlike   D. unlikely

Answers: (1)_____ (2)_____ (3)_____ (4)_____ (5)_____

2. Because it is so much faster than its __(1)__, a computer typically spends far more time waiting for input and output __(2)__ it does processing data. During a single second, a large __(3)__ can execute a million instructions or more, so each unused second represents a __(4)__ waste of potential computing power.

(1) A. peripherals   B. periodicals   C. periphrases   D. peripherics
(2) A. while   B. than   C. as   D. rather than
(3) A. server   B. computer   C. controller   D. mainframe
(4) A. huge   B. little   C. much   D. tremendous

Answers: (1)_____ (2)_____ (3)_____ (4)_____

3. In Excel, one of the most popular __(1)__ programs, the files used are called workbooks. These consist of __(2)__, which are the electronic form of large sheets of paper. The data is displayed in rows and __(3)__, in a large table on the screen. Cells are the small boxes formed where rows and columns intersect. Every cell has a(n) __(4)__, which contains the labels of the row and column which intersect to form it, for example D5. Cells can contain numeric data, __(5)__, formulas or functions. Text is used for titles, or to describe the figures being displayed. Numeric data can be entered using the keyboard or can result from calculations.

A. address   B. text   C. workbooks   D. worksheets
E. spreadsheet   F. rows   G. column

Answers: (1)_____ (2)_____ (3)_____ (4)_____ (5)_____

Ex4. Translate the following sentences into English.

1. 应用软件可分为两类：通用应用软件和专用应用软件。通用应用软件包括浏览器、字处理器、电子表格、数据库管理系统和演示图形。专用应用软件包括多媒体、图形、虚拟现实和人工智能程序。

2. 为了使浏览器连接其他资源，必须指定资源的位置或地址，这些地址称为 URL。浏览器一旦连接到网站，一个文档文件就会发送到你的计算机上，浏览器解释其中的 HTML 命令并将该文档显示为一个网页。通常网站的第一个页面称为首页。

3. 许多操作系统和应用程序都使用窗口来显示信息或请求输入，在计算机屏幕上可以同时打开并显示多个窗口。窗口通常可以改变大小、移动和关闭。

4. 数据和信息存储在工作簿中，工作簿包含一张或多张相关的工作表，列用字母标识，行用数字标识，一行一列的交叉部分产生一个单元格，在单元格内输入信息。

5. 数据库是相关数据的集合，DBMS 被用于创建和使用 DB。一个关系 DB 将数据组织成相关的表，每张表由称为记录的行和称为字段的列构成。

6. 套装软件是提供字处理、电子表格和数据库管理程序等功能的一个单独程序。

7. 文字处理软件让你能够创建、编辑、保存和打印基于文本的文档，如信件和报告。

Ex5. Translate the following sentences into Chinese.

1. File not found.
2. Bad command or file name.
3. Invalid drive specification.
4. File cannot be copied onto itself.
5. Invalid path , not directory , or directory not empty.
6. Delete one or more files.
7. Create a directory.
8. Rename a file or files.
9. Display or set the system time.
10. Display the contents of a text file.

Ex6. Speaking.

1. Make a list of the most important features offered by word processors.
2. What is a spreadsheet? What is it used for?
3. What will happen if you change the value of a cell of spreadsheet?
4. Explain the differences between general-purpose and special-purpose applications.

Part II   Reading materials

Apple Computer

A personal computer company founded in 1976 by Steven Jobs and Steve Wozniak.

Throughout the history of personal computing, Apple has been one of the most innovative influences. In fact, some analysts say that the entire evolution of the PC can be viewed as an effort to catch up with the Apple Macintosh.

In addition to inventing new technologies, Apple also has often been the first to bring sophisticated technologies to the personal computer. Apple's innovations include:

Graphical user interface (GUI). First introduced in 1983 on its Lisa computer. Many components of the Macintosh GUI have become de facto standards and can be found in other operating systems, such as Microsoft Windows.

Color. The Apple Ⅱ, introduced in 1977, was the first personal computer to offer color monitors.

Built-in networking. In 1985, Apple released a new version of the Macintosh with built-in support for networking (LocalTalk).

Plug & play expansion. In 1987, the Mac Ⅱ introduced a new expansion bus called NuBus that made it possible to add devices and configure them entirely with software.

QuickTime. In 1991, Apple introduced QuickTime, a multi-platform standard for video, sound, and other multimedia applications.

Integrated television. In 1993, Apple released the Macintosh TV, the first personal computer with built-in television and stereo CD.

RISC. In 1994, Apple introduced the Power Mac, based on the PowerPC RISC microprocessor.

术语参考译文

应用（应用软件）
软件应用程序是为最终用户设计的一个或一组程序。应用程序可以是系统软件或应用软件。

图形用户界面
图形用户界面（GUI）利用计算机的图形功能，使程序更容易使用点击设备、菜单和图标。

桌面
在图形用户界面中，桌面是用来描述文件系统的隐喻。这样的桌面由图片（称为图标）构成，图标可表示文件、文件夹和各种类型的文档（即信件、报告、图片）。

集成开发环境
简称 IDE，是集成到一个软件应用的编程环境，提供一个图形用户界面生成器、文本或代码编辑器、编译器/解释器和调试器。Visual Studio、Delphi、JBuilder、FrontPage 和 Dreamweaver 都是 IDE 的例子。

菜单
可以从菜单中选择命令或选项的列表。

对话框

一个在显示屏上显示信息或要求输入的框。

上下文菜单

当使用应用程序或操作系统时，单击两个按钮鼠标的右键（也称右击）出现的菜单。单键鼠标用户可以通过在单击的同时按住 Ctrl 键来弹出上下文菜单。

对齐

（1）在文本中，文本或图形相对于边缘的排列。

（2）关于图形对象，对齐描述它们的相对位置。

粗（黑）体字

一种比普通字体更暗的字体。大多数文字处理软件允许标记文本为粗体。

字体

一组字符的设计。字体是字样和其他特性的组合，如大小和间距。

边距

在字处理中，边距是指纸边缘的空白条。大多数文字处理软件允许指定边距的宽度。

自动换行

在字处理技术中，一种使文字处理软件强制所有文本位于限定的边距内的特性。

缩进

缩进用于描述将一个段落与左、右边距分开的距离，或空格的数量。

默认值

如果没有指定替代物，则设备或程序自动选择的值或设置称为默认值。例如，文字处理软件具有默认边距和默认页长度，可以重写或重置。

样式表

在字处理和桌面排版系统中，样式表是定义文档布局的文件或表单.。当填写样式表时，会指定页面大小、边距和字体等参数。

左对齐

沿左边距对齐文本。左对齐文本与刷新左文本（flush-left）相同。

图像尺寸

图像尺寸是数字图像的长度和宽度。它通常是以像素来衡量的，但是一些图形程序允许用等值的英寸或厘米来查看和处理图像。

跑马灯

在网页上，一个滚动的文本区域。从版本 2 开始，微软 Internet Explorer 支持特殊的 <MARQUEE> 标签以创建这些区域。

过滤器

一种程序，它接收某种类型的数据作为输入，以某种方式转换它，然后输出转换后的数据。

色彩饱和度

在图形和成像中，色彩饱和度被用来描述图像中颜色的强度。一个饱和的图像有太亮的颜色。

Unit 3   Operating Systems

Part I   Computer terms

- **Operating Systems**

The *operating system* (**OS**) is the most important program that runs on a computer. Every general-purpose computer must have an operating system to run other programs and applications. Computer operating systems perform basic tasks, such as recognizing input from the keyboard, sending output to the display screen, keeping track of files and directories on the disk, and controlling peripheral devices such as printers. Computers and mobile devices must have an operating system to run programs. (ref. Fig.1-3-1)

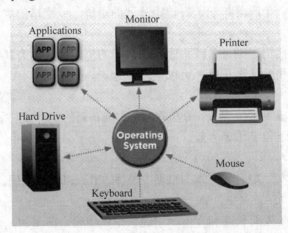

Fig.1-3-1   Operating system

For large systems, the operating system has even greater responsibilities and powers. It is like a traffic cop—it makes sure that different programs and users running at the same time do not interfere with each other. The operating system is also responsible for *security*, ensuring that unauthorized users do not access the system.

- **Multitasking**

The ability to execute more than one task at the same time, a task being a program. The terms multitasking and multiprocessing are often used interchangeably, although multiprocessing mplies that more than one CPU is involved.

In multitasking, only one CPU is involved, but it switches from one program to another so quickly that it gives the appearance of executing all of the programs at the same time.

- **Multiprocessing**

Supports running a program on more than one CPU.

- **Multi-user**

Refers to computer systems that support two or more simultaneous users. All mainframes and minicomputers are multi-user systems, but most personal computers and workstations are not.

- **Multithreading**

The ability of an operating system to execute different parts of a program, called *threads*, simultaneously. The programmer must carefully design the program in such a way that all the threads can run at the same time without interfering with each other.

- **Real time**

Responds to input instantly. General-purpose operating systems, such as DOS and UNIX, are not real-time.

- **Linux Operating Systems**

Linux is a freely distributed open source operating system that runs on a number of hardware platforms. The Linux kernel was developed mainly by Linus Torvalds and it is based on UNIX.

- **Windows Operating Systems**

Microsoft Windows is a family of operating systems for personal and business computers. Windows dominates the personal computer world, offering a graphical user interface (GUI), virtual memory management, multitasking, and support for many peripheral devices.

- **Mac Operating Systems**

Mac OS is the official name of the Apple Macintosh operating system. Mac OS features a graphical user interface (GUI) that utilizes windows, icons, and all applications that run on a Macintosh computer have a similar user interface.

- **Control panel**

A Macintosh utility that permits you to set many of the system parameters. For example, you can control the type of beeps the Mac makes and the sensitivity of the mouse.

- **Mobile operating system**

A mobile operating system, also called a mobile OS, is an operating system that is specifically designed to run on mobile devices such as mobile phones, smartphones, PDAs, tablet computers and other handheld devices. The mobile operating system is the software platform on top of which other programs, called application programs, can run on mobile devices.

- **Smartphone**

Smartphones are a handheld device that integrates mobile phone capabilities with the more common features of a handheld computer or PDA. Smartphones allow users to store information, e-mail, install programs, along with using a mobile phone in one device. For example a smartphone could be a mobile phone with some PDA functions integrated into the device, or vise versa.

- **NOS—network operating system**

Abbreviated as *NOS*, a network operating system includes special functions for connecting computers and devices into a local-area network (LAN). Some operating systems, such as UNIX and the Mac OS, have networking functions built in.

New Words

general-purpose	[ˈdʒenrəlˈpɜːpəs]	*adj.* 多方面的；通用的
traffic cop	[ˈtræfik kɔp]	*n.* 交通警察
interfere	[ˌintəˈfiə(r)]	*vi.* 干预；干涉；打扰
security	[siˈkjuərəti]	*n.* 安全；保证
unauthorized	[ʌnˈɔːθəraizd]	*adj.* 未授权的；未经许可的
execute	[ˈeksikjuːt]	*vt.* 执行；完成
simultaneous	[ˌsimlˈteiniəs]	*adj.* 同时的
mainframe	[ˈmeinfreim]	*n.* 大型机；主机
minicomputer	[ˈminikəmpjuːtə(r)]	*n.* 小型计算机
thread	[θred]	*n.* 线；线程
distribute	[diˈstribjuːt]	*vt.* 分配；散布；分发
dominate	[ˈdɔmineit]	*v.* 支配；影响
control panel	[kənˈtrəul ˈpænəl]	*n.* 控制面板
utility	[juːˈtiləti]	*n.* 实用程序；功用；效用
beep	[biːp]	*n.* 哔哔声 *v.* 嘟嘟响
sensitivity	[ˌsensəˈtivəti]	*n.* 敏感；灵敏性
mobile	[ˈməubail]	*adj.* 可移动的 *n.* 手机
tablet	[ˈtæblət]	*n.* 药片；平板电脑
install	[inˈstɔːl]	*vt.* 安装；安置

Phrases

mobile device	移动设备
real time	实时

handheld computer	手持计算机
traffic cop	交通警察
personal computer	个人计算机
control panel	控制面板
built in	内置的
system parameter	系统参数
mobile phone	移动电话
vise versa	反之亦然

Abbreviation

OS—Operating System　　　　　操作系统
NOS—Network Operating System　　网络操作系统
PDA—Personal Digital Assistant　　个人数字助理
LAN—Local Area Network　　　　局域网

Exercises

Ex1. Multiple choice.

1. The (　　) operating system is designed to run on Intel and Intel-compatible microprocessors.
　　A. Windows　　　B. Mac OS　　　C. Linux　　D. Sherlock　　E. Norton
2. (　　) is permanently stored in the computer and provides a link between hardware and other programs that run on the PC.
　　A. Interface　　B. Operating System　　C. Internet　　D. Application Software
3. The (　　) has several major components, including the system kernel, a memory management system, the file system manager, device drivers, and the system libraries.
　　A. application　　B. information system　　C. network　　D. OS
4. On a (　　) memory system, the logical memory space available to the program is totally independent of the physical memory space.
　　A. cache　　　B. virtual　　　　C. RAM　　　D. ROM
5. When you first turn power　(1)　, the portion of the operating system that is in ROM is in　(2)　of the PC. It is programmed to　(3)　the memory for malfunctions,　(4)　if everything is OK, to load　(5)　the remainder of the basic operating system.
　　(1) A. to　　　　B. on　　　　C. get　　　　D. put
　　(2) A. control　　B. run　　　　C. surface　　D. buffer
　　(3) A. reduce　　B. product　　　C. switch　　D. check
　　(4) (5) A. off　　B. for　　　　C. if　　　　D. in
　　　　E. on　　　　F. and

6. MS Windows is a __(1)__ that makes it easier to interact with your computer. These displayed elements are defined as follows:

__(2)__: A small picture used to represent a system component or command on the screen.

__(3)__: A list of available commands in an application windows.

__(4)__: A windows in which you can enter information what Windows or an application program needs to continue processing a command.

The design of Windows is based on the __(5)__ that your computer should be like your desktop. With Windows, you can keep many different files and projects available at the same time and __(6)__ among them just by reaching for the one you need—as if they __(7)__ on your desk.

    A. were   B. icon   C. dialog box   D. concept
    E. switch   F. graphical user interface   G. menu

Answers: (1) ____ (2) _____ (3) _____ (4) _____ (5) _____ (6) _____ (7) ____

Ex2. Match each numbered item with the most closely related lettered item.

    a. networking operating system   b. operating systems   c. icons
    d. mainframe computers   e. antivirus program   f. smartphone
    g. file compression program   h. backup software   i. multitasking
    j. embedded operating system

    (   ) 1. Programs that makes copies of files to be used if originals are lost or damaged.

    (   ) 2. Programs that coordinate computer resources.

    (   ) 3. Computers used primarily by large organizations for critical applications, bulk data processing, such as census, industry and consumer statistics, enterprise resource planning, and transaction processing.

    (   ) 4. A mobile phone with an advanced mobile operating system that combines features of a personal computer operating system with other features useful for mobile or handheld use.

    (   ) 5. Program that reduce the size of files for efficient storage.

    (   ) 6. Graphic elements that represent commonly used features.

    (   ) 7. A computer's ability to run more than one application at a time.

    (   ) 8. Operating systems completely stored within ROM memory.

    (   ) 9. Operating system used to control and coordinate computers that are linked together.

    (   ) 10. Program that guards your computer system against damaging and invasive programs..

Ex3. Computer English test.

1．Operating systems have developed over the past thirty years for two main purposes.

First, they provide a convenient environment for the development and execution of __(1)__. Second, operating systems attempt to __(2)__ computational activities to ensure good performance of the computing system. Initially, computers were used from the front __(3)__. Software such as assemblers, loaders, and compilers improved on the convenience of programming the system, but also required substantial set up time. To reduce the set up time, operators were hired and similar jobs were __(4)__ together.

(1) A. internets   B. devices   C. programs   D. computers
(2) A. schedule   B. adjust   C. arrange   D. plan
(3) A. desktop   B. console   C. controller   D. keyboard
(4) A. bonded   B. executed   C. finished   D. batched

Answers: (1) _____ (2) _____ (3) _____ (4) _____

2. ① Use __(1)__ expressions to initialize static and extern __(2)__.
② When the computer has been __(3)__, you must reboot operating system __(4)__ with a __(5)__ diskette.

(1)~(5) A. clear   B. fixed   C. infected   D. up   E. variables   F. installed
            G. new   H. again   I. constant   J. functions

Answers: (1) _____ (2) _____ (3) _____ (4) _____ (5) _____

Ex4. Translate the following sentences into English.

1. 语言翻译程序将程序员编写的程序指令转换成计算机能理解和处理的语言。

2. 系统软件由操作系统、实用程序、设备驱动程序和语言翻译程序构成。操作系统完成三个基本的功能：管理资源，提供用户接口以及运行应用程序。

3. 桌面是 Windows 提供的用户界面，Windows 将信息存储在文件和文件夹中。图标用于与 Windows 操作系统交互。另一种设计运行在 Macintosh 计算机上功能强大、易于使用的操作系统是 Mac OS。

4. 设备驱动程序是一种与操作系统共同工作以使得硬件设备与计算机系统通信的专门程序。

Ex5. Speaking.

1. What is operating system?
2. What do the terms means in your language: multithreading, multitasking, multiprocessing, multi-user?
3. How do you run a program on a computer with a graphic interface?
4. Can you give two reasons of the importance of user-friendly interfaces?
5. What can you use smartphone to do everyday?
6. What mainstream mobile operating system are there nowadays?

Part II   Reading materials

Smartphone

A smartphone is a mobile phone (also known as cell phones) with an advanced mobile operating system which combines features of a personal computer operating system with other features useful for mobile or handheld use. Smartphones, which are usually pocket-sized, typically combine the features of a mobile phone, such as the abilities to place and receive voice calls and create and receive text messages, with those of other popular digital mobile devices like personal digital assistants (PDAs), such as an event calendar, media player, video games, GPS navigation, digital camera and digital video camera. Most smartphones can access the Internet and can run a variety of third-party software components ("apps"). They typically have a color display with a graphical user interface that covers 70% or more of the front surface. The display is often a touchscreen, which enables the user to use a virtual keyboard to type words and numbers and press onscreen icons to activate "app" features.

In 1999, the Japanese firm NTT DoCoMo released the first smartphones to achieve mass adoption within a country. Smartphones became widespread in the late 2000s. Most of those produced from 2012 onward have high-speed mobile broadband 4G LTE, motion sensors, and mobile payment features. In the third quarter of 2012, one billion smartphones were in use worldwide. Global smartphone sales surpassed the sales figures for regular cell phones in early 2013. As of 2013, 65% of mobile consumers in the United States owned smartphones. By January 2016, smartphones held over 79% of the U.S. mobile market.

术语参考译文

操作系统
操作系统（OS）是在计算机上运行的最重要的程序。每一个通用计算机都必须有一个操作系统来运行其他程序和应用程序。计算机操作系统执行基本任务，如识别键盘输入、将输出发送到显示屏幕、跟踪磁盘上的文件和目录以及控制外围设备例如打印机。计算机和移动设备必须有一个操作系统来运行程序。

对于大型系统，操作系统有更大的责任和权力。它就像一个交通警察——它确保不同的程序和用户在同一时间运行而不互相干扰。操作系统还负责安全，确保未经授权的用户不能访问系统。

多任务处理
同时执行多个任务的能力，一个任务是一个程序。术语多任务处理和多重处理通常是可互换使用的，但多重处理意味着多个 CPU 的参与。

在多任务处理中，只涉及一个 CPU，但它会快速从一个程序切换到另一个程序，给人同时执行所有程序的感觉。

多重处理
支持在一个以上的处理器上运行一个程序。

多用户
指支持两个或多个同时用户的计算机系统。所有的大型机和小型机都是多用户系统，但大多数的个人计算机和工作站不是多用户的。

多线程
操作系统同时执行程序的不同部分（称为线程）的能力。程序员必须按如下方式仔细设计程序，即所有线程都可以同时运行而不会相互干扰。

实时

立即响应输入。通用操作系统，如 DOS 和 UNIX，不是实时的。

Linux 操作系统

Linux 是一个自由分发的开源操作系统，它运行在多个硬件平台上。Linux 内核主要是由 Linus Torvalds 开发的，它基于 UNIX。

Windows 操作系统

微软 Windows 是一个个人和商业电脑操作系统的家族。Windows 在个人电脑世界占主导地位，提供了图形用户界面（GUI）、虚拟内存管理、多任务处理，并支持许多外围设备。

Mac 操作系统

Mac OS 是苹果的 Macintosh 操作系统的正式名称。Mac OS 具有图形用户界面（GUI），利用窗口、图标，所有运行在 Macintosh 计算机上的应用程序有相似的用户界面。

控制面板

Macintosh 的实用工具，允许设置许多系统参数。例如，可以控制 Mac 发出的嘟嘟声的类型以及鼠标的灵敏度。

移动操作系统

移动操作系统，也被称为移动 OS，是专门设计运行在移动设备上的操作系统，如手机、智能手机、掌上电脑、平板电脑和其他手持设备。在移动操作系统这个软件平台上，其他应用程序才可以在移动设备上运行。

智能手机

智能手机是一个手持设备，集成了手机功能和掌上电脑或 PDA 的较常见功能。智能手机允许用户在一个设备上存储信息、收发电子邮件、安装程序，以及使用移动电话。例如，智能手机可以是集成到设备的具有某些 PDA 功能的移动电话，反之亦然。

NOS——网络操作系统

网络操作系统，简称 NOS，包括将电脑和设备与局域网（LAN）相连的特殊功能。一些操作系统，如 UNIX 和 Mac OS，有一些内置的网络功能。

Unit 4　　　　　　　　　　Data

Part I　Computer terms

- **Data compression**

Data compression refers to storing data in a format that requires less space. Data compression deals with file compression as well as lossless compression.

- **ASCII**

Abbreviated from American Standard Code for Information Interchange, is a character encoding standard. ASCII codes represent text in computers, telecommunications equipment, and other devices. Most modern character-encoding schemes are based on ASCII, although they support many additional characters.

- **Text file**

A file that holds text. The term *text file* is often used as a synonym for ASCII file, a file in which characters are represented by their ASCII codes.

- **Data structure**

In programming, the term *data structure* refers to a scheme for organizing related pieces of information. The basic types of data structures include:

　　files, lists, arrays, records, trees, tables

Each of these basic structures has many variations and allows different operations to be performed on the data.

- **Table**

Refers to data arranged in rows and columns. A spreadsheet, for example, is a table.

- **Record**

A complete set of information. Records are composed of fields, each of which contains one item of information. A set of records constitutes a file.

- **Linked list**

In computer science, a linked list is a linear collection of data elements, called nodes, each pointing to the next node by means of a pointer. It is a data structure consisting of a group of nodes which together represent a sequence. Under the simplest form, each node is composed of data and a reference (in other words, a link) to the next node in the sequence. This structure allows for efficient insertion or removal of elements from any position in the sequence during iteration. More complex variants add additional links, allowing efficient insertion or removal from arbitrary element references.

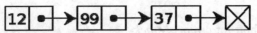

Linked lists are among the simplest and most common data structures. They can be used to implement several other common abstract data types, including lists (the abstract data type), stacks, queues, associative arrays, and S-expressions.

- **Array**

In programming, a series of objects all of which are the same size and type. Each object in an array is called an *array element*. For example, you could have an array of integers or an array of characters or an array of anything that has a defined data type. The important characteristics of an array are:

Each element has the same data type (although they may have different values).

The entire array is stored contiguously in memory (that is, there are no gaps between elements).

Arrays can have more than one dimension. A one-dimensional array is called a *vector* ; a two-dimensional array is called a *matrix*.

- **Tree structure**

A type of data structure in which each element is attached to one or more elements directly *beneath* it. The connections between elements are called branches.

The elements at the very bottom of an inverted tree (that is, those that have no elements below them) are called *leaves*. Inverted trees are the data structures used to represent hierarchical file structures. In this case, the leaves are files and the other elements above the leaves are directories.

A binary *tree* is a special type of inverted tree in which each element has only two branches below it.

- **Binary tree**

A special type of tree structure in which each node has at most two leaves. Binary tree are often used for sorting data, as in a heap sort.

- **Matrix**

(1) A two-dimensional array; that is, an array of rows and columns.

(2) The background area of color display.

- **Queue**

In programming, a queue is a data structure in which elements are removed in the same order they were entered. This is often referred to as FIFO (first in, first out). In contrast, a *stack* is a data structure (last in, first out).

- **Stack**

(1) In programming, a special type of data structure in which elements are removed in the reverse order from which they were entered. This is referred to as LIFO, so the most recently added item is the first one removed. This is also called *last-in, first-out (LIFO)*.

Adding an item to a stack is called *pushing*. Removing an item from a stack is called *popping*.

(2) In networking, short for *protocol stack*.

- **Vector**

(1) In computer programming, a one-dimensional array. A vector can also mean a pointer.

(2) In computer graphics, a line that is defined by its start and end point.

New Words

compression	[kəmˈpreʃn]	n. 压缩；压紧
telecommunication	[ˌtelikəˌmjuːniˈkeiʃn]	n. 电信
synonym	[ˈsinənim]	n. 同义词
array	[əˈrei]	n. 数组；队列；阵列
variation	[ˌveəriˈeiʃn]	n. 变化，变动
linear	[ˈliniə(r)]	adj. 直线的；[数]一次的；线性的
node	[nəud]	n. 结点
sequence	[ˈsiːkwəns]	n. 顺序；[数]数列，序列
reference	[ˈrefrəns]	n. 参考（书）；提及 v. 引用；参照
variant	[ˈveəriənt]	n.（词等的）变体；变量
insertion	[inˈsɜːʃn]	n. 插入（物）
removal	[riˈmuːvl]	n. 除去；搬迁
arbitrary	[ˈɑːbitrəri]	adj. 随意的，任性的
implement	[ˈimpliment]	vt. 实施，执行 n. 工具，器械
queue	[kjuː]	n. 队列
stack	[stæk]	n. 堆栈，垛
associative	[əˈsəuʃiətiv]	adj. 联合的，联想的
vector	[ˈvektə(r)]	n. 矢量，向量
matrix	[ˈmeitriks]	n. 矩阵；模型
beneath	[biˈniːθ]	prep. 在……的下方
invert	[inˈvɜːt]	vt. 使……前后倒置；使反转
hierarchical	[ˌhaiəˈrɑːkikl]	adj. 分层的；等级（制度）的
remove	[riˈmuːv]	vt. 去除；开除

Phrases

data compression	数据压缩
lossless compression	无损压缩

telecommunications equipment	电信设备
data structure	数据结构
abbreviated from	是…的缩写
synonym for	是…的同义词
linked list	链表
associative array	关联数组
array element	数组元素
one-dimensional array	一维数组
two-dimensional array	二维数组
tree structure	树结构
inverted tree	倒置树
hierarchical file structure	层次文件结构
binary tree	二叉树
protocol stack	协议栈
heap sort	堆排序
be referred to as	被称为
short for	是…的简称（简记）

Abbreviation

ASCII—American Standard Code for Information Interchange 美国信息交换标准码
FIFO—first-in, first-out 先进先出
LIFO—last-in, first-out 后进先出

Exercises

Ex1. Multiple choice.

1. (　　) is not a linear structure.
    A. Graph　　　　B. Queue　　　　C. Stack　　　　D. 1-dimension array

2. In (　　), the only element that can be deleted or removed is the one that was inserted most recently.
    A. a line　　　　B. a queue　　　　C. an array　　　　D. a stack

3. (　　): A collection of related information, organized for easy retrieval.
    A. Data　　　　B. Database　　　　C. Buffer　　　　D. Stack

4. The maximum number of data that can be expressed by 8 bits is (　　).
    A. 64　　　　B. 128　　　　C. 255　　　　D. 256

5. GIF files are limited to a maximum of 8 bits/pixel, it simply means that no more than 256 colors are allowed in (   ).

    A. an image   B. a file   C. a window   D. a page

6. A query is used to search through the database to locate a particular record or records, which conform to specified (   ).

    A. criteria   B. standards   C. methods   D. conditions

7. A DB represents a collection of information organized in such a way that a computer program can quickly (1) desired pieces of data. You can think of a DB as an electronic filing system. DB are generally organized by fields, records, and files. A (2) is a single piece of information; a (3) is one complete set of fields; and a (4) is a collection of records. For example, a telephone book is analogous to a file. It contains a list of records, each of which (5) of three fields : name, address, and telephone number.

To (6) information from a DB, you need a DBMS. This is a collection of programs that enables you to enter, organize, and select data in a DB.

    A. access   B. select   C. record   D. field
    E. file   F. consists

Answers: (1)_____ (2)_____ (3)_____ (4)_____ (5)_____ (6)_____

Ex2. Match each numbered item with the most closely related lettered item.

    a. query   b. stack   c. queue   d. array   e. linked list   f. table   g. binary tree

(   ) 1. A particular kind of abstract data type or collection in which the first element added to the queue will be the first one to be removed.

(   ) 2. A linear collection of data elements, called nodes, each pointing to the next node by means of a pointer. It is a data structure consisting of a group of nodes which together represent a sequence.

(   ) 3. A tree data structure in which each node has at most two children, which are referred to as the left child and the right child.

(   ) 4. An abstract data type and data structure based on the principle of Last In First Out (LIFO).

(   ) 5. A data structure consisting of a collection of elements (values or variables), each identified by at least one array index or key.

(   ) 6. A collection of related data held in a structured format within a database. It consists of columns, and rows.

(   ) 7. A way of searching for information in a database.

Ex3. Computer English Test.

1. An instruction is made up of operations that (1) the function to be performed and operands that represent the data to be operated on. For example, if an instruction is to perform the operation of (2) two numbers, it must know (3) the two numbers are. The processor's

job is to (4) instructions and operands from memory and to perform each operation .Having done that, it signals memory to send it (5) instruction.

(1) A. skip   B. smile   C. smoke   D. specify
(2) A. add   B. added   C. adding   D. addition
(3) A. when   B. where   C. which   D. who
(4) A. get   B. make   C. push   D. put
(5) A. ant   B. last   C. next   D. second

Answers: (1)_____ (2) _____ (3) _____ (4) _____ (5) _____

2. The arithmetic/logic unit(ALU) is the functional unit that (1) the computer with logical and computational capabilities. (2) are brought into the ALU by the (3) , and the ALU performs whatever arithmetic or logic operations are required to help carry out the instruction. (4) operations include adding, subtracting, multiplying, and dividing. Logic operations make a comparison and take action base on the results. For example, two numbers might be compared (5) determine if they are equal. If they are equal, processing will continue; if they are not equal, processing will stop.

(1) A. provides   B. provided   C. provide   D. is providing
(2) A. Codes   B. data   C. Bytes   D. Bits
(3) A. logic unit B. control unit C. arithmetic/logic unit D. arithmetic unit
(4) A. Control   B. Logic   C. Arithmetic   D. Arithmetic/logic
(5) A. with   B. of   C. to   D. for

Answers: (1)_____ (2) _____ (3) _____ (4) _____ (5) _____

Ex4. Translate the following sentences into English.

1．因特网是全球最大的计算机网络，WWW 是一种因特网服务，它提供访问因特网上可用资源的多媒体接口。

2．微机硬件由各种不同的设备构成。有三种类型的设备：主机包括 CPU 和内存，输入输出设备是人与计算机间的翻译者，辅存设备存储数据和程序。常用的介质包括软盘、硬盘和光盘。

3．数据通常以电子形式保存在文件中，常见的文件类型有文档文件、电子表格文件、数据库文件和演示文件。

4．主要有两种类型的软件：系统软件和应用软件。用户与应用软件交互，系统软件使得应用软件和计算机硬件交互。

5．软件是程序的另一个名称，程序由告诉计算机如何完成其工作的一步一步的指令构成。软件的目的是将数据转换成信息。

Ex5. Speaking.

1. How is western character represented in a computer?
2. How to define and initialize an one-demensional integer array in C language?

3. Explain the terms with your language: array, linked list, queue, stack.
4. What basic types of data structures are there?

Part II   Reading materials

Data structure

In computer science, a data structure is a particular way of organizing data in a computer so that it can be used efficiently. Data structures can implement one or more particular abstract data types(ADT), which specify the operations that can be performed on a data structure and the computational complexity of those operations. In comparison, a data structure is a concrete implementation of the specification provided by an ADT.

Different kinds of data structures are suited to different kinds of applications, and some are highly specialized to specific tasks. For example, relational databases commonly use B-tree indexes for data retrieval, while compiler implementations usually use hash tables to look up identifiers.

Data structures provide a means to manage large amounts of data efficiently for uses such as large databases and internet indexing services. Usually, efficient data structures are key to designing efficient algorithms. Some formal design methods and programming languages emphasize data structures, rather than algorithms, as the key organizing factor in software design. Data structures can be used to organize the storage and retrieval of information stored in both main memory and secondary memory.

There are numerous types of data structures, generally built upon simpler primitive data types:
- An array is a number of elements in a specific order, typically all of the same type. Elements are accessed using an integer index to specify which element is required (Depending on the language, individual elements may either all be forced to be the same type, or may be of almost any type). Typical implementations allocate contiguous memory words for the elements of arrays (but this is not always a necessity). Arrays may be fixed-length or resizable.
- A linked list (also just called list) is a linear collection of data elements of any type, called nodes, where each node has itself a value, and points to the next node in the linked list. The principal advantage of a linked list over an array, is that values can always be efficiently inserted and removed without relocating the rest of the list. Certain other operations, such as random access to a certain element, are however slower on lists than on arrays.
- A record (also called tuple or struct) is an aggregate data structure. A record is a value that contains other values, typically in fixed number and sequence and typically indexed by names. The elements of records are usually called fields or members.

- A union is a data structure that specifies which of a number of permitted primitive types may be stored in its instances, e.g. float or long integer. Contrast with a record, which could be defined to contain a float and an integer; whereas in a union, there is only one value at a time. Enough space is allocated to contain the widest member data type.
- A tagged union (also called variant, variant record, discriminated union, or disjoint union) contains an additional field indicating its current type, for enhanced type safety.
- A class is a data structure that contains data fields, like a record, as well as various methods which operate on the contents of the record. In the context of object-oriented programming, records are known as plain old data structures to distinguish them from classes.

术语参考译文

数据压缩

数据压缩是指按照需要较少空间的格式存储数据。数据压缩涉及文件压缩以及无损压缩。

ASCII

美国信息交换标准代码的缩写，是一种字符编码标准。ASCII 码在计算机、远程通信设备和其他设备上表示文本。大多数现代的字符编码方案都基于 ASCII，虽然它们支持许多额外的字符。

文本文件

保存文本的文件。术语文本文件通常被用作 ASCII 文件的同义词，ASCII 文件中的字符是用 ASCII 码表示的。

数据结构

在编程中，术语数据结构是指组织相关信息的方案。数据结构的基本类型包括：
文件、列表、数组、记录、树、表。
每一个这些基本结构都有许多变化，并且允许对数据执行不同的操作。

表

表指按行和列排列的数据。例如，一个电子表格是一个表。

记录

一系列完整的信息。记录由字段组成，每个字段包含一个信息项。一组记录构成一个文件。

链表

在计算机科学中，链表是数据元素的线性集合，这些数据元素称为结点，每个结点通过指针指向下一个结点。它是由一组结点构成的数据结构，这些结点共同代表一个序列。在最简单的形式下，每个结点由数据和一个指向序列中下一个结点的引用（换句话说，链接）构成。这种结构允许在迭代过程中从序列中的任何位置有效地插入或删除元素。更复杂的形式是添加额外的链接，允许从任意元素位置高效地插入或删除元素。

链表是最简单和最常见的数据结构之一。它们可以用来实现其他几种常见的抽象数据类型，包括列表（抽象数据类型）、栈、队列、关联数组和 S 表达式。

数组

在编程中，指一系列有相同大小和类型的对象。数组中的每个对象称为数组元素。例如，你可以拥有一个整数数组或一个字符数组或任何已定义数据类型的数组。数组的重要特征是：

- 每个元素具有相同的数据类型（虽然它们可能有不同的值）。
- 整个数组存储在连续内存中（即元素之间没有间隙）。

数组可以有一个以上的维度。一维数组称为向量；二维数组称为矩阵。

树结构

一种数据结构，其中每个元素之下直接连接有一个或多个元素。元素之间的连接称为分支。

倒置树的底部的元素（也就是说，它们下面没有元素）叫作叶子。倒置树是用来表示分层文件结构的数据结构。在这种情况下，叶子是文件，叶子上面的其他元素是目录。

二叉树是一种特殊类型的倒置树，其中每个元素在它下面只有两个分支。

二叉树

一种特殊类型的树结构，其中每个结点至多有两片叶子。二叉树通常用于排序数据，如堆排序。

矩阵

（1）一个二维数组，也就是由行和列构成的数组。

（2）彩色显示器的背景区域。

队列

在程序设计中，队列是一种数据结构，其中元素被删除的顺序与元素进入的顺序相同。这通常被称为 FIFO（先进先出）。与之相反，堆栈是一种"后进先出"的数据结构。

栈

（1）在程序设计中，指一种特殊类型的数据结构，其中元素被删除的顺序与它们进入的顺序相反。这被称为 LIFO，即最近添加的项目是第一个被删除的项目。这也被称为后进先出（LIFO）。

向栈中添加项称为进栈。从栈中移除项称为退栈。

（2）在网络中，是协议栈的缩写。

向量

（1）在计算机编程中，指一维数组。一个向量也可以意味着一个指针。

（2）在计算机图形学中，指由起始点和结束点定义的一条线。

Unit 5　Programming

Part I　Computer terms

- **Compile**

To transform a program written in a high-level programming language from source code into object code. Programmers write programs in a form called source code.

- **Compiler**

A program that translates source code into object code.

- **Object code**

The code produced by a compiler. The source code consists of instructions in a particular language, like C or FORTRAN. Computers, however, can only execute instructions written in a low-level language called *machine language*.

- **Expression**

In programming, an expression is any legal combination of symbols that represents a value. Each programming language and application has its own rules for what is legal and illegal.

Every expression consists of at least one operand and can have one or more operators. Operands are values, whereas operators are symbols that represent particular actions.

- **Identifier**

Same as *name*. The term *identifier* is usually used for variable names.

- **Subscript**

(1) In programming, a symbol or number used to identify an element in an array. Usually, the subscript is placed in brackets following the array name. For example, AR[5] identifies element number 5 in an array called AR.

(2) In word processing, a character that appears slightly below the line, as in this example: H_{20}. A *superscript* is a character that appears slightly above the line.

- **OOP**

Object-**o**riented **p**rogramming (**OOP**) refers to a type of computer programming (software design) in which programmers define not only the data type of a data structure, but also the types of operations (functions) that can be applied to the data structure.

In this way, the data structure becomes an object that includes both data and functions. In addition, programmers can create relationships between one object and another. For example, objects can inherit characteristics from other objects.

- **Programming language**

A vocabulary and set of grammatical rules for instructing a computer to perform specific tasks. The term programming language usually refers to high-level languages, such as BASIC, C, C++, COBOL, FORTRAN, Ada, and Pascal. Each language has a unique set of keywords and a special syntax for organizing program instructions.

High-level programming languages, while simple compared to human languages, are more complex than the languages the computer actually understands, called machine languages. Each different type of CPU has its own unique machine language.

- **Function**

(1) In programming, a named section of a program that performs a specific task. In this sense, a function is a type of procedure or routine. Some programming languages make a distinction between a *function*, which returns a value, and a *procedure*, which performs some operation but does not return a value.

Most programming languages come with a prewritten set of functions that are kept in a library. You can also write your own functions to perform specialized tasks.

(2) The term *function* is also used synonymously with *operation* and command. For example, you execute the delete *function* to erase a word.

- **Procedure**

(1) Same as routine, subroutine, and function. A procedure is a section of a program that performs a specific task.

(2) An ordered set of tasks for performing some action.

- **Parameter**

(1) Characteristic. For example, *specifying parameters* means defining the characteristics of something. In general, parameters are used to customize a program. For example, filenames, page lengths, and font specifications could all be considered parameters.

(2) In programming, the term *parameter* is synonymous with argument, a value that is passed to a routine.

- **HTML—HyperText Markup Language**

Short for *HyperText Markup Language,* the authoring language used to create documents on the World Wide Web. HTML is similar to SGML, although it is not a strict subset.

HTML defines the structure and layout of a Web document by using a variety of tags and attributes. The correct structure for an HTML document starts with <HTML><HEAD><BODY> and ends with </BODY></HTML>. All the information you'd like to include in your Web page fits in between the <BODY> and </BODY> tags.

- **XML**

Short for *Extensible Markup Language,* a specification developed by the W3C. XML is a pared-down version of SGML, designed especially for Web documents. It allows designers to create their own customized tags, enabling the definition, transmission, validation, and interpretation of data between applications and between organizations.

New Words

compile	[kəmˈpail]	vt. 编译；编制；汇编
compiler	[kəmˈpailə(r)]	n. 汇编者；编译程序
expression	[ikˈspreʃn]	n. 表现，表示，表达式
symbol	[ˈsimbl]	n. 符号；象征；标志
legal	[ˈli:gl]	adj. 法律的；合法的
illegal	[iˈli:gl]	adj. 非法的，违法的
operator	[ˈɔpəreitə(r)]	n. 操作员；运算符
operand	[ˈɔpərænd]	n. 操作数；运算数
variable	[ˈveəriəbl]	n. 可变因素，变量　adj. 变化的
identifier	[aiˈdentifaiə(r)]	n. 标识符
subscript	[ˈsʌbskript]	n. 下标，脚注
identify	[aiˈdentifai]	vt. 确定；识别，认出
function	[ˈfʌŋkʃn]	n. 功能，作用；函数
object	[ˈɔbdʒikt]	n. 物体；目标；对象
inherit	[inˈherit]	vt. & vi. 继承
characteristic	[ˌkærəktəˈristik]	n. 特性，特征，特色
unique	[juˈni:k]	adj. 仅有的；独一无二的
procedure	[prəˈsi:dʒə(r)]	n. 程序，手续；过程，步骤
routine	[ru:ˈti:n]	n. 常规；例行程序
library	[ˈlaibrəri]	n. 库；图书馆
synonymous	[siˈnɔniməs]	adj. 同义词的；同义的
customize	[ˈkʌstəmaiz]	vt. 定制，定做；
operation	[ˌɔpəˈreiʃn]	n. 手术；操作；[数]运算
command	[kəˈmɑ:nd]	n. 命令，指挥；指令
delete	[diˈli:t]	vt. & vi. 删除

erase	[iˈreiz]	vt. 清除；擦掉
subroutine	[ˈsʌbruːtiːn]	n. 子程序
document	[ˈdɔkjumənt]	n. 文档，证件
specification	[ˌspesifiˈkeiʃn]	n. 规格；说明书；详述
pared-down	[ˈpeədd'aun]	adj. 压缩的；简化的
attribute	[əˈtribjuːt]	vt. 认为……是；把……归于 n. 属性；特征
author	[ˈɔːθə(r)]	n. 作者 vt. 创作出版
definition	[ˌdefiˈniʃn]	n. 定义；规定；[物]清晰度
transmission	[trænsˈmiʃn]	n. 传送；播送
validation	[ˌvæliˈdeiʃn]	n. 确认
interpretation	[inˌtɜːpriˈteiʃn]	n. 理解；解释，说明

Phrases

high-level programming language	高级编程语言
source code	源代码
object code	目标代码
machine language	机器语言
low-level language	低级语言
variable names	变量名
font specifications	字体规范
is similar to	与……类似
Web documents	Web 文档
customized tag	定制标签
pared-down version	精简版
make a distinction between…and…	对……和……加以区别

Abbreviation

OOP—Object-oriented programming　　　　面向对象程序设计
HTML—HyperText Markup Language　　　　超文本标记语言
WWW—World Wide Web　　　　万维网
SGML—Standard for General Markup Language　　　　通用标记语言标准
XML—Extensible Markup Language　　　　可扩展标记语言

Exercises

Ex1. Multiple choice.

1. C++ is used with proper (   ) design techniques.
    A. object oriented   B. object based   C. fact to object   D. face to target
2. (   ) is not a programming language.
    A. COBOL   B. Java   C. UML   D. BASIC
3. In C language, when an array name is passed to a function what is passed is the (   ) of the beginning of the array.
    A. data   B. value   C. location   D. element
4. In C language, the result of the logical (   ) operator is 1 if the value of its operand is 0, and is 0 if the value of its operand is non-zero.
    A. AND   B. NOT   C. OR   D. EOR
5. Very long, complex expressions in program are difficult to write correctly and difficult to (   ).
    A. defend   B. detect   C. default   D. debug
6. The standard (   ) in C language contain many useful functions for input and output, string handling, mathematical computations, and system programming tasks.
    A. database   B. files   C. libraries   D. subroutine
7. The term "(   ) program" means a program written in high-level language.
    A. compiler   B. executable   C. source   D. object
8. A programmer must know about a function's (   ) to call it correctly.
    A. location   B. algorithm   C. interface   D. statements
9. In (   ) programming, the programmer determines the sequence of instructions to be executed, not the user.
    A. top-down   B. structure   C. data-driven   D. event-driven
10. In C language, functions are important because they provide a way to (   ) code so that a large complex program can be written by combining many smaller parts.
    A. modify   B. modularize   C. block   D. board

Ex2. Match each numbered item with the most closely related lettered item.

a. computer language   b. optimize   c. program   d. bug   e. compile
f. modify   g. boot   h. byte   i. portable   j. parallel

(   ) 1. Carried or moved with ease.

(   ) 2. The amount of computer memory needed to store one character of a specified size, usually 8 bits for a microcomputer and 16 bits for a large computer.

(   ) 3. Of or relating to the simultaneous transmission of all the bits of a byte over spate wires.

(   ) 4. A system of symbols and rules used for communication with or between computers.

( ) 5. To make as perfect or effective as possible.
( ) 6. A defect in the code or routine of a program.
( ) 7. To translate (a program) into machine language.
( ) 8. To start an operating system.
( ) 9. An instruction sequence in programed instruction.
( ) 10. To change in form or character; alter.

Ex3. Computer English test.

1. The programs for the earliest digital computers were written in a machine language. Pure machine language programming required the programmer to write out the sequences of __(1)__ or decimal digits by which each instruction was represented in the computer memory. By the mid 1950s it was realized that programmers could specify instruction codes and memory locations by __(2)__ memonics, which could be translated into the internal machine language by a translation program called an __(3)__.

In the late 1950s and in the 1960s, procedure oriented language were developed to allow programmers to specify algorithms in a notation natural to the problem being solved. Programs specified in a procedure oriented language were translated into the internal language of a particular computer by a translation program called a __(4)__. The commonly used programming languages in the 1960s and 1970s included Fortran, Algol 60, Coblo, PL1, and APL.

(1) A. binary   B. bytes   C. marks   D. instructions
(2) A. signal   B. symbolic   C. graphic   D. symbol
(3) A. assembler   B. operator   C. interpreter   D. executer
(4) A. super mailer   B. translator   C. compiler   D. real player

Answers: (1)_____ (2)_____ (3)_____ (4)_____

2. In __(1)__ to solve a computational problem, its solution must be __(2)__ in terms of a sequence of computational steps, each of which may be effectively performed by a human agent or by a digital computer. Systematic notations for the specification of such sequence are referred to as programming __(3)__. A specification of the sequence in a particular programming language is referred to as a program. The task of developing programs for the solution of computational problems is referred to as __(4)__. A person engaging in the activity of programming is referred to as a __(5)__.

(1) A. order   B. ordered   C. ordering   D. orders
(2) A. specify   B. specified   C. specifies   D. specifying
(3) A. instructions   B. commands   C. notes   D. languages
(4) A. program   B. programs   C. programming   D. programmed
(5) A. professor   B. professors   C. programmer   D. programmers

Answers: (1)_____ (2)_____ (3)_____ (4)_____ (5)_____

Ex4. Translation the following sentences into English.

1. 备份程序复制文件以备万一原始文件丢失或损坏时使用。

2. Windows 操作系统自带了一些实用程序，包括备份、磁盘清理和磁盘碎片整理。

3. Cache 用作内存与 CPU 之间的临时高速存储区域。Cache 用于保存 RAM 中最常访问的信息。

4. 通常被称为 ALU 的算术逻辑单元，完成两类运算：算术和逻辑运算。算术运算即基本的数学运算：加、减、乘、除。逻辑运算由比较单元构成。

5. RAM 芯片被称为临时或易失的，是因为如果断电其内容会丢失。

Ex5. Translate the following sentences into Chinese.

1. File not found.
2. Bad command or file name.
3. Invalid drive specification.
4. File cannot be copied onto itself.
5. Invalid path, not directory or directory not empty.

Ex6. Speaking.

1. What is programming? What programming languages have you learned?
2. What is HTML?
3. What is the rule to name a identifier in C language?
4. Explain how a program is produced.

Part II   Reading materials

The Basics: Object Oriented Programming Concepts

If you are new to object-oriented programming languages, you will need to know a few basics before you can get started with code.

- Abstraction: The process of picking out (abstracting) common features of objects and procedures.
- Class: A category of objects. The class defines all the common properties of the different objects that belong to it.
- Encapsulation: The process of combining elements to create a new entity. A procedure is a type of encapsulation because it combines a series of computer instructions.
- Information hiding: The process of hiding details of an object or function. Information hiding is a powerful programming technique because it reduces complexity.
- Inheritance: a feature that represents the "is a" relationship between different classes.
- Interface: the languages and codes that the applications use to communicate with each other and with the hardware.
- Messaging: Message passing is a form of communication used in parallel programming and object-oriented programming.
- Object: a self-contained entity that consists of both data and procedures to manipulate the data.

- Polymorphism: A programming language's ability to process objects differently depending on their data type or class.
- Procedure: a section of a program that performs a specific task.

Advantages of Object Oriented Programming

One of the principal advantages of object-oriented programming techniques over procedural programming techniques is that they enable programmers to create modules that do not need to be changed when a new type of object is added. A programmer can simply create a new object that inherits many of its features from existing objects. This makes object-oriented programs easier to modify.

术语参考译文

编译

将高级编程语言编写的程序从源代码转换为目标代码。程序员以一种称为源代码的形式编写程序。

编译器

将源代码转换为目标代码的程序。

目标代码

由编译器产生的代码。源代码由特定语言的指令构成，如 C 或 Fortran。然而，计算机只能执行用低级语言（称为机器语言）编写的指令。

表达式

在编程中，表达式是符号的任何合法组合，表达式代表一个值。每个编程语言和应用程序都有自己的规则，指明什么是合法的和非法的。

每个表达式由至少一个操作数组成，并且可以有一个或多个操作符。操作数是值，而运算符是表示特定动作的符号。

标识符

与名字相同。术语标识符通常用于变量名。

下标

（1）在程序设计中，用来识别数组中元素的符号或数字。通常，下标放置在数组名称后面的括号中。例如，AR[5]识别 AR 数组中的 5 号元素。

（2）在文字处理、显示略低于线的字符，如下面的例子：H_{20}。上标是略高于线的字符。

面向对象的程序设计

面向对象编程（OOP）是一种计算机编程（软件设计），程序员定义的不仅是数据结构的数据类型，还有可应用于数据结构操作的类型（功能）。

这样，数据结构就变成了包含数据和功能的对象。此外，程序员可以创建一个对象和另一个对象之间的关系。例如，对象可以继承来自其他对象的特性。

编程语言

指导计算机执行特定任务的词汇和语法规则。编程语言通常是指高级语言，如 Basic、C、C++、Cobol、Fortran、Ada 和 Pascal。每种语言都有一套独特的关键词和特殊的语法来组织程序指令。

高级编程语言虽然比人类语言简单，但比计算机实际理解的语言复杂得多，被称为机器语言。每个不同类型的 CPU 都有自己独特的机器语言。

函数

（1）在编程过程中，执行特定任务的一个命名的程序部分。在这个意义上，函数是一种过程或例程。一些编程语言分为有返回值的函数和执行一些操作但不返回值的过程。

大多数编程语言附带一组预先写好的函数，保存在库里。也可以编写自己的函数来执行特定的任务。

（2）术语函数也是操作和命令的同义词。例如，执行删除函数来删除一个单词。

过程

（1）与例程、子例程和函数相同。过程是执行特定任务的程序的一部分。

（2）执行某些动作的一组有序任务。

参数

（1）特征。例如，指定参数意味着定义某事物的特性。通常，参数用于自定义程序。例如，文件名、页面长度和字体规格都可以被认为是参数。

（2）在编程中，术语参数是变元的同义词，是传递给例程的值。

超文本标记语言 HTML

超文本标记语言（HTML）是用于在万维网上创建文档的创作语言。HTML 与 SGML 一样，虽然它不是一个严格的子集。

HTML 通过使用各种标记和属性，定义了一个 Web 文档的结构和布局，一个 HTML 文档正确的结构以<HTML><HEAD><BODY>开始，以</BODY></HTML>结束。你想在网页中包含的所有信息都位于<BODY>和</BODY>标签之间。

XML

可扩展标记语言，由 W3C 制定的规范。XML 是 SGML 的一个精简版，专为网页文件设计。它允许设计人员创建自己定制的标签，使应用程序之间和组织之间能够定义、传输、验证和解释数据。

Unit 6　　Computer Science

Part I　Computer terms

- **Artificial Intelligence**

Artificial Intelligence (AI) refers to a branch of computer science that is focused on making computers behave in human-like instances. Artificial intelligence plays a part in computer games, simulations and even online chat bots.

- **Handwriting recognition**

The technique by which a computer system can recognize characters and other symbols written by hand in natural handwriting. The technology is used for identification and also on devices such as PDA and tablet PCs where a stylus is used to handwrite on a screen with a stylus, after which the computer turns the handwriting into digital text.

- **Pattern recognition**

An important field of computer science concerned with recognizing patterns, particularly visual and sound patterns. It is central to optical character recognition (OCR), voice recognition, and handwriting recognition.

- **Voice recognition**

The field of computer science that deals with designing computer systems that can recognize spoken words. Note that voice recognition implies only that the computer can take dictation, not that it *understands* what is being said. Comprehending human languages falls under a different field of computer science called *natural language processing*.

- **Natural language processing**

Natural language processing (NLP) is the analyzing, understanding and generating the languages that humans use naturally in order to interface with computers.

- **Face recognition**

A type of biometrics that uses images of a person's face for recognition and identification.

- **Machine learning**

In computer science machine learning refers to a type of data analysis that uses algorithms that learn from data. It is a type of artificial intelligence (AI) that provides systems with the ability to learn without being explicitly programmed. This enables computers to find data within data without human intervention.

What is important to know about machine learning is that data is being used to make predictions, not code. Data is dynamic so machine learning allows the system to learn and evolve with experience and the more data that is analyzed.

- **Data mining**

A class of database applications that look for hidden patterns in a group of data that can be used to predict future behavior.

- **Neural network**

A type of artificial intelligence that attempts to imitate the way a human brain works. Rather than using a digital model, in which all computations manipulate zeros and ones, a neural network works by creating connections between *processing elements*, the computer equivalent of neurons. The organization and weights of the connections determine the output.

- **Biometrics**

Biometrics refers to the techniques for authentication which rely on measurable characteristics that can be automatically checked. Different aspects of biometrics include biometric data, matching and verification.

- **Client/Server Computing**

Client/Server computing is a computing model in which client and server computers communicate with each other over a network. In client/server computing, a server takes requests from client computers and shares its resources, applications and/or data with one or more client computers on the network, and a client is a computing device that initiates contact with a server in order to make use of a shareable resource.

- **Databases**

Databases, or DBs, are collections of information or data organized so that computer applications known as database management systems (DBMS) can quickly and efficiently access specific data from the database. The data in traditional databases are organized by fields, records, and files。

- **Record**

A complete set of information. Records are composed of fields, each of which contains one item of information. A set of records constitutes a file.

- **Field**

A space allocated for a particular item of information. In database systems, fields are the smallest units of information you can access.

- **Distributed computing**

Distributed Computing refers to a type of computing where different components and object making up an application can be located on different computers that are connected to a network.

- **Cloud computing**

Cloud computing is a type of computing that relies on *sharing computing resources* rather than having local servers or personal devices to handle applications.

- **Measurement**

Measurement refers to the size, length, or amount of something defined through the action of measuring. In computing we use measurements to describe things like volumes of data or CPU speeds.

- **Nanotechnology**

Nanotechnology is a branch of engineering that works to design and create extremely small electronic circuits as well as mechanical devices. This involves dealing with manipulating at the level of single atoms and molecules.

- **Pervasive computing**

Refers to the idea that technology is moving further than just the personal computer into everyday devices. Pervasive computing takes into consideration nanotechnology as well as smart cards.

- **Supercomputing**

Supercomputing refers to the fastest type of computer which takes into consideration cloud storage, cloud computing, and nanotechnology.

- **Parallel processing**

The simultaneous use of more than one CPU to execute a program. Parallel processing makes a program run faster because there are more CPUs running it.

- **High-performance computing**

A branch of computer science that concentrates on developing supercomputers and software to run on supercomputers. A main area of this discipline is developing parallel processing algorithms and software: programs that can be divided into little pieces so that each piece can be executed simultaneously by separate processors.

- **Transaction processing**

Refers to a type of computer processing where a computer responds to a user's request immediately. Transaction processing includes eBusiness as well as transaction authority markup language.

- **Virtualization**

Virtualization is the creation of a virtual version of a device or resource—such as a server or a network—where the framework divides the resource into one or more environments. Virtualization includes emulation as well as physical topology.

New Words

simulation	[ˌsimjuˈleiʃn]	n. 模仿，模拟
natural	[ˈnætʃrəl]	adj. 自然的；天生的

handwriting	[ˈhændraitiŋ]	n. 书法；笔迹
stylus	[ˈstailəs]	n. 唱针；尖笔
pattern	[ˈpætn]	n. 模式；图案；花样
analyze	[ˈænəlaiz]	vt. 分析；分解
recognition	[ˌrekəgˈniʃn]	n. 认识，识别
identification	[aiˌdentifiˈkeiʃn]	n. 认同；鉴定，识别；验明
explicitly	[ikˈsplisitli]	adv. 明白地，明确地
algorithm	[ˈælgəriðəm]	n. 算法；运算法则
prediction	[priˈdikʃn]	n. 预言，预测
mining	[ˈmainiŋ]	n. 采矿；挖掘
behavior	[biˈheivjə]	n. 行为；态度
neural	[ˈnjuərəl]	adj. 神经的
manipulate	[məˈnipjuleit]	vt. 操纵；操作，处理
biometrics	[ˌbaiəuˈmetriks]	n. 生物识别技术
client	[ˈklaiənt]	n. 顾客；客户机
record	[ˈrekɔ:d]	n. 唱片；记录
field	[fi:ld]	n. 字段；场地，领域
measurement	[ˈmeʒəmənt]	n. 量度；尺寸
volume	[ˈvɔlju:m]	n. 体积；卷；音量
molecule	[ˈmɔlikju:l]	n. 分子
atom	[ˈætəm]	n. 原子；原子能
simultaneously	[ˌsiməlˈteiniəsli]	adv. 同时地；一齐
parallel	[ˈpærəlel]	adj. 平行的；并行的
supercomputer	[ˈsu:pəkəmpju:tə(r)]	n. 超级计算机，巨型计算机
transaction	[trænˈzækʃn]	n. 交易，业务，事务
authority	[ɔːˈθɔrəti]	n. 权威；权力
virtualization	[vɜːtʃuəlaiˈzeiʃn]	n. 虚拟化
emulation	[ˌemjuˈleiʃn]	n. 竞赛；仿效
version	[ˈvɜːʃn]	n. 版本；译文
environment	[inˈvairənmənt]	n. 环境，外界

Phrases

computer science	计算机科学
handwriting recognition	手写识别
pattern recognition	模式识别
voice recognition	语音识别
face recognition	人脸识别
data analysis	数据分析
human intervention	人为干预
make prediction	做出预测
data mining	数据挖掘
database application	数据库应用
neural network	神经网络
digital model	数字模型
biometric data	生物数据
matching and verification	匹配和验证
Client/Server Computing	客户机/服务器计算
communicate with	与……通信
distributed computing	分布式计算
cloud computing	云计算
electronic circuit	电子电路
mechanical device	机械装置
atom and molecule	原子与分子
pervasive computing	普适计算
smart card	智能卡
cloud storage	云存储
high-performance computing	高性能计算
parallel processing algorithm	并行处理算法
transaction processing	交易处理
physical topology	物理拓扑

Abbreviation

AI—Artificial Intelligence　　　　　　　　人工智能

OCR—optical character recognition　　光学字符识别
NLP—Natural language processing　　自然语言处理

Exercises

Ex1. Multiple Choice.

Please (1) new diskette for drive A.
And press any key when (2) .
Press ESC key to (3) to main menu.
Please specify the (4) and the filename.
Please (5) the new volume label.
(1)~(5): A. enter　　B. memory　　C. ready　　D. interface　　E. return
　　　　F. insert　　G. exit　　H. device　　I. install　　J. path
Answers: (1)_____ (2)_____ (3)_____ (4)_____ (5)_____

Ex2. Match each numbered item with the most closely related lettered item.

a. client-server model　　b. algorithm　　c. network topology　　d. DBMS
e. parallel computing　　f. speech recognition

(　) 1. It is the inter-disciplinary sub-field of computational linguistics that develops methodologies and technologies that enables the recognition and translation of spoken language into text by computers.

(　) 2. A computer software application that interacts with the user, other applications, and the database itself to capture and analyze data, which is designed to allow the definition, creation, querying, update, and administration of databases.

(　) 3. A type of computation in which many calculations or the execution of processes are carried out simultaneously.

(　) 4. The arrangement of the various elements (links, nodes, etc.) of a computer network.

(　) 5. A distributed application structure that partitions tasks or workloads between the providers of a resource or service, called servers, and service requesters, called clients.

(　) 6. A self-contained sequence of actions to be performed, which performs calculation, data processing, and/or automated reasoning tasks.

Ex3. Computer English Test.

1. Artificial Intelligence (1) a relatively young branch of science, new enough that we can still trace the development of the field from its inception in 1956 to the present. About six years ago, when we (2) planning the Handbook of Artificial Intelligence, we thought it would (3) possible to present AI comprehensively in three volumes. In retrospect, that seems to have (4) a good guess, although, inevitably, the outline has (5) changed many times to reflect changes in the emphasis and methods of AI.

(1)~(5): A. be B. am C. are D. is
E. was F. were G. being H. been

Answers: (1) _____ (2) _____ (3) _____ (4) _____ (5) _____

2. The obvious advantage of a Graphical User Inter face(GUI) is to organize the computer to make __(1)__ from a human __(2)__ , rather than to force users to adapt to the __(3)__ of computers and software. GUI is converging to the point where a __(4)__ person can walk up to a computer, experiment briefly with the mouse and the __(5)__ objects on screen, and gain some understanding of how to accomplish basic tasks.

(1)~(5): A. iconic B. converge C. sense D. use
E. perspective F. capable G. peculiarities H. imaging

Answers: (1) _____ (2) _____ (3) _____ (4) _____ (5) _____

3. Software quality assurance is now an __(1)__ sub-discipline of software engineering. As Buckly and Oston point out, __(2)__ software quality assurance is likely to lead to an ultimate __(3)__ of software costs. However, the major hurdle in the path of software management in this area is the lack of __(4)__ software standards. The development of accepted and generally applicable standards should be one of the principal goals of __(5)__ in software engineering.

(1) A. emerging B. emergent C. engaging D. evolve
(2) A. effective B. effortless C. light D. week
(3) A. balance B. growth C. production D. reduction
(4) A. usable B. usage C. useless D. useness
(5) A. management B. planning C. production D. research

Answers: (1) _____ (2) _____ (3) _____ (4) _____ (5) _____

Ex4. Translate the following sentences into English.

1. 终端是将你连接到大型机或被称为主机或服务器的其他类型的计算机上的一个输入输出设备。

2. 最常见的定点设备是鼠标，三种基本的鼠标设计包括机械鼠标、光学鼠标和无线鼠标。类似于鼠标的设备包括跟踪球、触摸屏和定点操纵杆。

3. 显示在显示器上的图像称为软拷贝，在纸上的信息输出称为硬拷贝。有几种类型的打印机：喷墨打印机、激光打印机、热敏打印机以及点阵打印机、连锁打印机和绘图仪等打印设备。

Ex5. Speaking.

1. What does the following terms mean, Artificial Intelligence, parallel computing, cloud computing?

2. What is a database? Which tasks can be performed by using a database? Make a list of possible applications.

3. What do the terms means in your language: file, record, field?

4. List common measurement unit of memory.

Part II Reading materials

Artificial Intelligence

Artificial intelligence (AI) is intelligence exhibited by machines. In computer science, an ideal "intelligent" machine is a flexible rational agent that perceives its environment and takes actions that maximize its chance of success at some goal. Colloquially, the term "artificial intelligence" is applied when a machine mimics "cognitive" functions that humans associate with other human minds, such as "learning" and "problem solving". As machines become increasingly capable, mental facilities once thought to require intelligence are removed from the definition. For example, optical character recognition is no longer perceived as an exemplar of "artificial intelligence", having become a routine technology. Capabilities currently classified as AI include successfully understanding human speech, competing at a high level in strategic game systems (such as Chess and Go), self-driving cars, and interpreting complex data. Some people also consider AI a danger to humanity if it progresses unabatedly. AI research is divided into subfields that focus on specific problems or on specificapproaches or on the use of a particular tool or towards satisfying particular applications.

The central problems (or goals) of AI research include reasoning, knowledge, planning, learning, natural language processing(communication),perception and the ability to move and manipulate objects. General intelligence is among the field's long-term goals. Approaches include statistical methods,computational intelligence, soft computing (e.g.machine learning), and traditional symbolic AI. Many tools are used in AI, including versions of search and mathematical optimization,logic,methods based on probability and economics. The AI field draws upon computer science,mathematics,psychology,linguistics,philosophy,neuroscience and artificial psychology.

The field was founded on the claim that human intelligence "can be so precisely described that a machine can be made to simulate it".This raises philosophical arguments about the nature of the mind and the ethics of creating artificial beings endowed with human-like intelligence, issues which have been explored by myth, fiction and philosophy since antiquity. Attempts to create artificial intelligence have experienced many setbacks, including the ALPAC report of 1966, the abandonment of perceptrons in 1970, the Lighthill Report of 1973, the second AI winter 1987—1993 and the collapse of the Lisp machine market in 1987. In the twenty-first century, AI techniques have become an essential part of the technology industry, helping to solve many challenging problems in computer science.

Reasoning, problem solving

Early researchers developed algorithms that imitated step-by-step reasoning that humans use when they solve puzzles or make logical deductions (reason). By the late 1980s and 1990s, AI research had developed methods for dealing with uncertain or incomplete information, employing concepts from probability and economics.

For difficult problems, algorithms can require enormous computational resources—most experience a "combinatorial explosion": the amount of memory or computer time required becomes astronomical for problems of a certain size. The search for more efficient problem-solving algorithms is a high priority.

Human beings ordinarily use fast, intuitive judgments rather than step-by-step deduction that early AI research was able to model. AI has progressed using "sub-symbolic" problem solving:embodied agent approaches emphasize the importance of sensorimotor skills to higher reasoning;neural net research attempts to simulate the structures inside the brain that give rise to this skill; statistical approaches to AI mimic the human ability.

Knowledge representation

Knowledge representation and knowledge engineering are central to AI research. Many of the problems machines are expected to solve will require extensive knowledge about the world. Among the things that AI needs to represent are: objects, properties, categories and relations between objects; situations, events, states and time; causes and effects; knowledge about knowledge (what we know about what other people know); and many other, less well researched domains. A representation of "what exists" is an ontology: the set of objects, relations, concepts and so on that the machine knows about. The most general are called upper ontologies, which attempt to provide a foundation for all other knowledge.

Planning

Intelligent agents must be able to set goals and achieve them. They need a way to visualize the future (they must have a representation of the state of the world and be able to make predictions about how their actions will change it) and be able to make choices that maximize the utility(or "value") of the available choices.

In classical planning problems, the agent can assume that it is the only thing acting on the world and it can be certain what the consequences of its actions may be. However, if the agent is not the only actor, it must periodically ascertain whether the world matches its predictions and it must change its plan as this becomes necessary, requiring the agent to reason under uncertainty.

Learning

Machine learning is the study of computer algorithms that improve automatically through experience and has been central to AI research since the field's inception.

Unsupervised learning is the ability to find patterns in a stream of input. Supervised learning includes both classification and numericalregression. Classification is used to determine what category something belongs in, after seeing a number of examples of things from several categories. Regression is the attempt to produce a function that describes the relationship between inputs and outputs and predicts how the outputs should change as the inputs change. In reinforcement learning the agent is rewarded for good responses and punished for bad ones. The agent uses this sequence of rewards and punishments to form a strategy for operating in its problem space. These three types of learning can be analyzed in terms of decision theory, using concepts like utility. The mathematical analysis of machine learning algorithms and their performance is a branch of theoretical computer science known as computational learning theory.

Within developmental robotics, developmental learning approaches were elaborated for lifelong cumulative acquisition of repertoires of novel skills by a robot, through autonomous self-exploration and social interaction with human teachers, and using guidance mechanisms such as active learning, maturation, motor synergies, and imitation.

Natural language processing (communication)

Natural language processing gives machines the ability to read and understand the languages that humans speak. A sufficiently powerful natural language processing system would enable natural language user interfaces and the acquisition of knowledge directly from human-written sources, such as newswire texts. Some straightforward applications of natural language processing include information retrieval, text mining, question answering and machine translation.

A common method of processing and extracting meaning from natural language is through semantic indexing. Increases in processing speeds and the drop in the cost of data storage makes indexing large volumes of abstractions of the user's input much more efficient.

Perception

Machine perception is the ability to use input from sensors (such as cameras, microphones, tactile sensors, sonar and others more exotic) to deduce aspects of the world. Computer vision is the ability to analyze visual input. A few selected subproblems are speech recognition, facial recognition and object recognition.

Motion and manipulation

The field of robotics is closely related to AI. Intelligence is required for robots to be able to handle such tasks as object manipulation and navigation, with sub-problems of localization (knowing where you are, or finding out where other things are), mapping (learning what is around you, building a map of the environment), and motion planning (figuring out how to get there) or path planning (going from one point in space to another point, which may involve compliant motion—where the robot moves while maintaining physical contact with an object).

术语参考译文

人工智能

人工智能（AI）是指计算机科学的一个分支，它致力于使计算机的表现与人类类似。人工智能在计算机游戏、模拟甚至在线聊天机器人方面发挥作用。

手写识别

计算机系统能识别手写字符和其他符号的技术。该技术用于识别，也用在如 PDA 和平板电脑这样的设备上，用手写笔在屏幕上写字，之后计算机将笔迹转成数字文本。

模式识别
计算机科学的一个重要领域，涉及识别模式，特别是视觉和声音模式。它是光学字符识别（OCR）、语音识别、手写识别的核心。

语音识别
计算机科学的一个领域，致力于设计能够识别口语单词的计算机系统。请注意，语音识别只意味着计算机可以听写，而不是它明白在说什么。理解人类语言属于不同的计算机科学领域，称为自然语言处理。

自然语言处理
为了与计算机交互，自然语言处理（NLP）是分析、理解和生成人类自然使用的语言。

人脸识别
一种生物识别技术，将人脸图像用作识别和身份认证。

机器学习
在计算机科学中，机器学习是指使用从数据中学习的算法进行数据分析。它是一种人工智能（AI），提供系统学习的能力，而不是被显式编程。这使得计算机能够在数据中找到数据而不需要人为干预。

关于机器学习，必须知道数据是被用做预测，而不是编码。数据是动态的，所以机器学习允许系统根据经验和更多的分析数据来学习和发展。

数据挖掘
一类数据库应用程序，在一组数据中寻找隐藏的模式，这些数据可以用来预测将来的行为。

神经网络
一种试图模仿人类大脑工作方式的人工智能。它不是使用数字模型，在数字模型中所有计算使用 0 和 1，神经网络按照在处理单元之间创建连接的方式工作，电脑相当于神经元。组织和连接的权重确定了输出。

生物识别技术
生物识别技术是指依赖于可自动检测的可测量特性的身份验证技术。生物识别的不同方面包括生物特征数据、匹配和验证。

客户机/服务器计算
客户机/服务器计算是一种计算模型，客户机和服务器计算机通过网络相互通信。在客户机/服务器计算中，服务器接收客户端计算机的请求，与一个或多个网络上的客户机共享资源、应用程序和/或数据，客户端是一个计算设备，它为了使用共享资源而启动与服务器的联系。

数据库
数据库是信息或数据的集合，这种组织方式使得称为数据库管理系统（DBMS）的计算机应用程序可以从数据库中快速有效地访问特定的数据。传统数据库中的数据由字段、记录和文件组成。

记录
一组完整的信息。记录由字段组成，每个字段包含一个信息项。一组记录构成一个文件。

字段
分配给一个特定信息项的空间。在数据库系统中，字段是可以访问的最小信息单元。

分布式计算
分布式计算是指将构成一个应用程序的不同组件和对象定位于网络不同计算机上的一种计算类型。

云计算
云计算是一种依赖于共享计算资源而不是本地服务器或个人设备来处理应用的计算类型。

测量
测量是指通过测量行为而定义的某物的大小、长度或数量。在计算中，我们使用测量来描述数据量或 CPU 速度等。

纳米技术
纳米技术是工程的一个分支，用于设计和制造非常小的电子电路以及机械设备。这涉及单个原子和分子级别的处理操作。

普适计算
普适计算的思想是指技术进一步转向每天的日常设备，而不仅仅是个人电脑。普适计算需要考虑纳米技术以及智能卡。

超级计算
超级计算是指最快的计算机类型，其考虑了云存储、云计算和纳米技术。

并行处理
同时使用多个 CPU 来执行一个程序。并行处理使程序运行得更快，因为有更多 CPU 运行它。

高性能计算
计算机科学的一个分支，致力于开发超级计算机和在超级计算机上运行的软件。这门学科的一个主要领域是开发并行处理算法和软件：可以分为小块的程序，使每一块可以被单独的处理器同时执行。

交易处理
交易处理指计算机直接响应用户请求的计算机处理类型。交易处理包括电子商务以及交易权限标记语言。

虚拟化
虚拟化是创建一个设备或资源的虚拟版本（如服务器或网络），框架将资源分割成一个或多个环境。虚拟化包括仿真和物理拓扑。

Unit 7 Multimedia

Part I Computer terms

● **Multimedia**

The use of computers to present text, graphics, video, animation, and sound in an integrated way. (ref. Fig.1-7-1)

Fig.1-7-1 Examples of individual content forms combined in multimedia

● **Multimedia kit**

A package of hardware and software that adds multimedia capabilities to a computer. Typically a multimedia kit includes a CD-ROM or DVD player, a sound card, speakers, and a bundle of CD-ROMs.

● **Animation**

A simulation of movement created by displaying a series of pictures, or frames. Cartoons on television is one example of animation. Animation on computers is one of the chief ingredients of multimedia presentations. There are many software applications that enable you to create animations that you can display on a computer monitor.

Note the difference between animation and video. Whereas video takes continuous motion and breaks it up into discrete frames, animation starts with independent pictures and puts them together to form the illusion of continuous motion.

- **Video**

(1) Refers to recording, manipulating, and displaying moving images, especially in a format that can be presented on a television.

(2) Refers to displaying images and text on a computer monitor.

- **Video conferencing**

Conducting a conference between two or more participants at different sites by using computer networks to transmit audio and video data.

- **Virtual reality**

An artificial environment created with computer hardware and software and presented to the user in such a way that it appears and feels like a real environment. To "enter" a virtual reality, a user dons special gloves, earphones, and goggles, all of which receive their input from the computer system. In this way, at least three of the five senses are controlled by the computer. In addition to feeding sensory input to the user, the devices also monitor the user's actions. The goggles, for example, track how the eyes move and respond accordingly by sending new video input.

- **Digital camera**

A digital camera stores images digitally rather than recording them on film. The pictures can be downloaded to a computer system.

- **Color depth**

The number of distinct colors that can be represented by a piece of hardware or software. Color depth is sometimes referred to as bit depth because it is directly related to the number of bits used for each pixel.

- **Convergence**

The coming together of two or more disparate disciplines or technologies. For example, the so-called fax revolution was produced by a convergence of telecommunications technology, optical scanning technology, and printing technology.

- **Page template**

A page template, or *Web page template*, often refers to a predesigned Web page that you can customize. The page template would include font, style, formatting, tables, graphics and other elements commonly found on a Web page.

- **Digital-to-analog converter**

Short for *digital-to-analog converter,* a device (usually a single chip) that converts digital data into analog signals. Modems require a DAC to convert data to analog signals that can be carried by telephone wires.

- **Analog-to-digital converter**

ADC is a device that converts analog signals into digital signals.

- **Modem**

A modem (modulator-demodulator) is a device or program that enables a computer to transmit data over, for example, telephone or cable lines.

Multimedia

- **Encryption**

The translation of data into a secret code. Encryption is the most effective way to achieve data security.

- **Decryption**

The process of decoding data that has been encrypted into a secret format. Decryption requires a secret key or password.

New Words

multimedia	[ˌmʌltiˈmiːdiə]	n. 多媒体
video	[ˈvidiəu]	n. 录像；录像磁带；录像机
audio	[ˈɔːdiəu]	adj. 音频的；听觉的
speaker	[ˈspiːkə(r)]	n. 扬声器
animation	[ˌæniˈmeiʃn]	n. 动画片
kit	[kit]	n. 成套用品；配套元件
package	[ˈpækidʒ]	vt. 包装　n. 包裹；包
bundle	[ˈbʌndl]	n. 捆；一批
frame	[freim]	n. 框架；边框
cartoon	[kɑːˈtuːn]	n. 漫画；动画片
ingredient	[inˈgriːdiənt]	n. 因素；组成部分；要素
presentation	[ˌpreznˈteiʃn]	n. 陈述，报告
discrete	[diˈskriːt]	adj. 分离的，离散的
illusion	[iˈluːʒn]	n. 错觉；幻想；假象
motion	[ˈməuʃn]	n. 运动；手势；动机
digital	[ˈdidʒitl]	adj. 数字的
analog	[ˈænəlɔːg]	n. 类似物；模拟　adj. 模拟的
participant	[pɑːˈtisipənt]	n. 参加者，参与者
earphone	[ˈiəfəun]	n. 耳机，听筒
goggle	[ˈgɔgl]	n. 护目镜；凝视
don	[dɔn]	vt. 穿上，披上
glove	[glʌv]	n. 手套
sensory	[ˈsensəri]	adj. 感觉的，感官的

feed	[fi:d]	vt. 喂养；向……提供
track	[træk]	vt. 跟踪；追踪
conference	[ˈkɔnfərəns]	n. 会议；讨论
depth	[depθ]	n. 深度；深处
discipline	[ˈdisəplin]	n. 纪律；学科；训练
template	[ˈtempleit]	n. 模板；样板
modulator	[ˈmɔdjuleitə]	n. 调制器；调节器
demodulator	[diːˈmɔdjuleitə]	n. 解调器；检波器
modem	[ˈməudem]	n. 调制解调器
converter	[kənˈvɜːtə(r)]	n. 变换器；变压器
encryption	[inˈkripʃn]	n. 加密；编密码
decryption	[diːˈkripʃn]	n. 解密，译码
decode	[ˌdiːˈkəud]	vt. 译（码），解（码）

Phrases

DVD player	DVD 播放机
moving image	动态图像
sound card	声卡
multimedia kit	多媒体套件
animation and video	动画与视频
continuous motion	连续运动
discrete frame	离散帧
computer monitor	计算机显示器
color depth	颜色深度
digital camera	数码相机
virtual reality	虚拟现实
video conferencing	视频会议
Web page template	网页模板
audio and video	音频与视频
3D surround sound	三维环绕音效
compression formation	压缩结构
analog signal	模拟信号

digital signal	数字信号
modulator-demodulator	调制器和解调器
telephone wire	电话线
cable line	电缆线
data security	数据安全

Abbreviation

DAC—Digital-to-Analog Converter　　　数模转换器
ADC—Analog-to-Digital Converter　　　模数转换器

Exercises

Ex1. Multiple choice.

1. The (　) controls a pointer on the screen. When you move it, the pointer moves too.
   A. menu　　　　B. icon　　　　C. mouse　　　　D. click

2. The type of installation "(　)" is selected by most users, the software will be installed with the most common options.
   A. Typical　　　B. Complete　　C. Custom　　　D. Minimum

3. When you choose a command name that is followed by '…' on menu, a (　) box appears in which you provide more information.
   A. text　　　　B. list　　　　C. check　　　　D. dialog

4. (　) is an online reference tool that you can use as you work.
   A. Title　　　　B. Edit　　　　C. Help　　　　D. View

5. You can cut, copy, and paste information quickly by clicking (　) bar buttons with the mouse.
   A. title　　　　B. tool　　　　C. status　　　　D. formula

6. To verify that your printer is operating correctly, you can run the printer (　) program.
   A. browse　　　B. self-test　　C. view　　　　D. driver

7. In general, you install software by the (　) program in CD-ROM.
   A. readme　　　B. option　　　C. load　　　　D. setup

8. You can doubleclick the program (　) to start the program.
   A. icon　　　　B. folder　　　C. graphic　　　D. menu

9. Sound cards, which allow recorded and synthesized sound to be __(1)__ by the computer, are often referred to as being 8-bit or 16-bit. Sound, which is analog, must be __(2)__ to digital data to be used by a computer. This is __(3)__ by sampling a sound wave at certain time intervals. 8-bit sound cards are usually capable of recording sound at 22.1 kHz, or 22,100 samples per second; 16-bit cards are capable of 44.2 kHz sampling, the sampling rate of music CDs.

Video, since it is a series of images, takes up an incredible amount of space. Computers are not capable of presenting full-screen digital images at the speed necessary for video. Therefore, this video must be __(4)__ in order for the computer to display it. The standard for video compression is MPEG, and MPEG devices are __(5)__, which allow computers to fully utilize MPEG-compressed videos.

A. available   B. played   C. accomplished   D. converted   E. compressed

Answers: (1) _____ (2) _____ (3) _____ (4) _____ (5) _____

Ex2. Match each numbered item with the most closely related lettered item.

a. website   b. animation   c. encryption   d. technological convergence
e. virtual reality   f. video   g. frame   h. graphic   i. digital signal
j. hyperlink

(   ) 1. An electronic medium for the recording, copying, playback, broadcasting, and display of moving visual media.

(   ) 2. A collection of related web pages, including multimedia content, typically identified with a common domain name, and published on at least one web server.

(   ) 3. A signal that represents a sequence of discrete values.

(   ) 4. The process of making the illusion of motion and the illusion of change by means of the rapid display of a sequence of images that minimally differ from each other.

(   ) 5. An element in an electronic document that links to another place in the same document or to an entirely different document.

(   ) 6. Refers to computer technologies that use software to generate the realistic images, sounds and other sensations that replicate a real environment (or create an imaginary setting), and simulate a user's physical presence in this environment.

(   ) 7. The tendency for different technological systems to evolve toward performing similar tasks.

(   ) 8. The pictorial representation and manipulation of data.

(   ) 9. The process of encoding messages or information in such a way that only authorized parties can access it.

(   ) 10. One of the many single photographic images in a motion picture.

Ex3. Computer English test.

1. If we break the word multimedia into its component parts, we get multi—meaning more than one, and media—meaning form of communication. Those types of media include: Text, Audio Sound, Static Graphics Images, __(1)__, and Full-Motion Video. Text is the basis for __(2)__ programs and is still the fundamental information used in many multimedia programs. In fact, many multimedia applications are __(3)__ on the conversion of a book to a computerized form. This conversion gives the user immediate access to the text and lets him or her display pop-up windows, which give definitions of certain words. As a multimedia programmer, you can choose what font to display text in, how big (or small) it should be, and what color it should be

displayed in. By __(4)__ text in more than one format, the message a multimedia application is trying to portray can be made more understandable. One type of application, which many people use every day, is the Windows Help Engine. This application is a text based information viewer that makes accessing information related __(5)__ a certain topic easy.

  (1) A. Sound   B. Animation   C. pictures   D. Dynamic Images
  (2) A. applications   B. word processing   C. communication   D. drawing
  (3) A. based   B. used   C. relied   D. depended
  (4) A. observing   B. searching   C. watching   D. displaying
  (5) A. with   B. on   C. to   D. in
  Answers: (1) _____ (2) _____ (3) _____ (4) _____ (5) _____

2. By using MP3, a 600MB music CD can be __(1)__ to 50MB or less. It can be streamed so that you can begin listening to the opening bars while the __(2)__ of the file arrives in the background. And, most important, MP3 music files retain good listening __(3)__ that __(4)__ compression schemes lacked. That __(5)__ of features makes accessing and distributing music on the Web practical for the first time.

  (1) A. pressed   B. compressed   C. moved   D. contained
  (2) A. past   B. next   C. rest   D. host
  (3) A. amount   B. mass   C. sense   D. quality
  (4) A. earlier   B. before   C. later   D. backward
  (5) A. addition   B. combination   C. difference   D. condition
  Answers: (1) _____ (2) _____ (3) _____ (4) _____ (5) _____

3. When most people refer to multimedia, they generally mean the combination of two or more continuous media, usually with some user __(1)__. In practice, the two media are normally audio and video, that is, __(2)__ plus moving __(3)__.

It should be obvious by now that transmitting multimedia material in uncompressed form is completely out of __(4)__. The only hope is that massive compression is possible. Fortunately, a large body of research over the past few decades has led to many compression techniques and algorithms that make multimedia transmission __(5)__.

  (1)~(3): A. display   B. games   C. help   D. interaction   E. pictures
        F. sound   G. web
  (4)~(5): A. impossible   B. fearful   C. feasible   D. program   E. thing
  Answers: (1) _____ (2) _____ (3) _____ (4) _____ (5) _____

4. On time-shared system, users access a computer through terminals. Some terminals are __(1)__, linked directly to a computer by cables, while others are __(2)__, communicating with distant computers over telephone lines or other transmission media.

Computers, however, do not store data as continuous waves; they store and manipulate discrete pulses. Because of this electronic incompatibility, whenever data are transmitted between a computer and remote terminal, they must be converted from pulse form to wave form, and back again. Converting to wave form is called __(3)__; converting back is called __(4)__; the task is performed by a hardware device called __(5)__. Normally, there is one at each end of a communication line.

(1)~(2): A. host   B. local   C. remote   D. client

(3)~(5): A. data set(modem)   B. data management   C. router   D. modulation

         E. demodulation   F. carrier signal

Answers: (1) _____ (2) _____ (3) _____ (4) _____ (5) _____

Ex4. Translate the following sentences into English.

1. 输出数据最常见的方式是通过键盘，键盘将人们理解的数字、字母和特殊字符转换成电子信号。

2. 数码相机类似于传统相机，除了图像是以数字方式记录在磁盘上或相机内存中而不是存在胶片上。

3. 语音输入设备将声音转换成能被计算机处理的格式。这些声音可来自各种各样的声源。迄今为止最广泛使用的语音输入设备是麦克风。

4. 最常使用的输出设备是显示器。显示器两个重要的特点是尺寸和清晰度。显示器的清晰度是用分辨率表示的，分辨率用像素来衡量，对于给定尺寸的显示器，分辨率越大，图像的清晰度越好。

Ex5. Speaking.

1. What is multimedia?

2. What is the advantage of computer multimedia presentations?

3. What is animation? What applications are used to make animation or cartoon?

4. What do the term virtual reality mean?

5. Can an existing PC can be adapted for multimedia applications?

Part II Reading materials

Big data

Big data is a term for data sets that are so large or complex that traditional data processing applications are inadequate to deal with them. Challenges include analysis, capture, data curation, search, sharing, storage, transfer, visualization, querying, updating and information privacy. The term "big data" often refers simply to the use of predictive analytics, user behavior analytics, or certain other advanced data analytics methods that extract value from data, and seldom to a particular size of data set. "There is little doubt that the quantities of data now available are indeed large, but that's not the most relevant characteristic of this new data ecosystem."

Analysis of data sets can find new correlations to "spot business trends, prevent diseases, combat crime and so on". Scientists, business executives, practitioners of medicine, advertising and governments alike regularly meet difficulties with large data-sets in areas including Internet search, finance, urban informatics, and business informatics. Scientists encounter limitations in e-Science work, including meteorology, genomics, connectomics, complex physics simulations, biology and environmental research.

Data sets grow rapidly—in part because they are increasingly gathered by cheap and numerous information-sensing mobile devices, aerial (remote sensing), software logs, cameras, microphones, radio-frequency identification (RFID) readers and wireless sensor networks. The world's technological per-capita capacity to store information has roughly doubled every 40 months since the 1980s; as of 2012, every day 2.5 exabytes (2.5×1018) of data is generated. One question for large enterprises is determining who should own big-data initiatives that affect the entire organization.

Relational database management systems and desktop statistics- and visualization-packages often have difficulty handling big data. The work may require "massively parallel software running on tens, hundreds, or even thousands of servers". What counts as "big data" varies depending on the capabilities of the users and their tools, and expanding capabilities make big data a moving target. "For some organizations, facing hundreds of gigabytes of data for the first time may trigger a need to reconsider data management options. For others, it may take tens or hundreds of terabytes before data size becomes a significant consideration."

术语参考译文

多媒体
使用计算机以集成的方式呈现文本、图形、视频、动画和声音。

多媒体套件
为计算机增加多媒体功能的一整套软硬件。一个典型的多媒体套件包括一个 CD-ROM 或 DVD 播放器、声卡、音箱和一组光盘。

动画
通过显示一系列图片或帧来创建运动的模拟。电视动画片是动画的一个例子。计算机动画是多媒体演示的主要内容之一。有很多的软件应用，可用于创建可以在计算机显示器上显示的动画。

注意动画和视频的区别。视频需要连续运动，并将其分割成离散帧，动画以独立的图片开始，并把它们放在一起形成连续运动的错觉。

视频
（1）记录、处理和显示运动图像，特别是用可以在电视上显示的格式。

（2）指在计算机显示器上显示的图像和文本。

视频会议
通过使用计算机网络传送音频和视频数据，在不同地点的两个或多个参与者之间进行会议。

虚拟现实
用计算机硬件和软件创造的人工环境，并以这样一种方式呈现给用户，它看起来和感觉起来都像真实的环境。要"进入"虚拟现实环境，用户要戴上特制的手套、耳机和护目镜，所有这些用于接收计算机系统的输入。这样，五种感官中至少有三种由计算机控制。

除了向用户提供感官输入，设备还监视用户的操作。例如，护目镜通过发送新的视频输入来跟踪眼睛的移动并做出相应响应。

数码相机
数码相机存储数字化图像，而不是将它们记录在胶片上。这些图片可以下载到计算机系统。

颜色深度
可以用一种硬件或软件表示的不同颜色的数目。颜色深度有时被称为位深度，因为它直接关系到每个像素所使用的位数。

汇聚
两个或多个不同学科或技术的融合。例如，所谓的传真革命，是由于电信技术、光扫描技术和打印技术的融合而产生的。

网页模板
一个页面模板，或网页模板，通常指的是一个预先设计好的网页，可以自定义。页面模板包括字体、样式、格式、表格、图形和网页上通常可找到的其他元素。

数模转换器
数字-模拟转换器的简称，是将数字数据转换为模拟信号的设备（通常是一个芯片）。调制解调器需要一个 DAC 将数据转换为可以在电话线上传输的模拟信号。

模数转换器
模数转换器 ADC 是一种将模拟信号转换成数字信号的设备。

调制解调器
调制解调器（调制器-解调器）是一种设备或程序，它使计算机能够在电话或电缆线路上传输数据。

加密
将数据翻译成一个秘密代码。加密是实现数据安全最有效的方式。

解密
将已加密为机密格式的数据解码的过程。解密需要密钥或密码。

Unit 8   Networks

Part I   Computer terms

- **Network**

A network is a group of two or more computer systems linked together. Computers on a network are sometimes called nodes. Computers and devices that allocate resources for a network are called servers.

- **Star network**

A local-area network (LAN) that uses a star topology in which all nodes are connected to a central computer.

- **Ring network**

A local-area network (LAN) whose topology is a ring. That is, all of the nodes are connected in a closed loop.

- **Bus network**

A network in which all nodes are connected to a single wire (the bus) that has two endpoints. Ethernet 10Base-2and 10Base-5 networks, for example, are bus networks.

- **Networking standards**

Networking standards ensure the interoperability of networking technologies by defining the rules of communication among networked devices. Networking standards exist to help ensure products of different vendors are able to work together in a network without risk of incompatibility.

- **Routing**

Routing is a phrase that refers to the process of moving a packet of data in a network from its source to a destination. Routing is usually performed by a dedicated networking device called a router and is a key feature of the Internet in that it enables messages to pass from one computer to another and eventually reach the target machine.

By forwarding network traffic among networks and selecting the best paths to utilize for moving data through a network, routing ensures optimal networking efficiency.

- **Hub**

A common connection point for devices in a network. Hubs serve as a central connection for all network equipment and handles a data type known as frames.

- **Router**

A device that forwards data packets along networks. A router is connected to at least two networks and are located at gateways.

- **Switch**

In networks, a device that filters and forwards packets between LAN segments. Switches operate at the data link layer (layer 2) and sometimes the network layer (layer 3) of the OSI Reference Model and therefore support any packet protocol.

- **Repeater**

A network device used to regenerate or replicate a signal used in transmission systems to regenerate signals distorted by transmission loss.

- **Application sharing**

A feature of many video conferencing applications that enables the conference participants to simultaneously run the same application.

- **Network protocols**

Network protocols are the language of rules and conventions used for handing communicated between network devices and ensuring the optimal operation of a network. Network protocols include key internet protocols such as IP and IPv6 as well as DNS and FTP, and it also includes more network-specific protocols like SNMP and NTP.

- **Network topologies**

Network topologies is a phrase that refers to the layout of connected devices on a network. How different nodes in a network are connected to each other and how they communicate are determined by the network's topology. Network topologies can apply to either physical or logical layout, and common type of network topologies include bus, ring, tree, star and mesh.

- **Collision detection**

Collision detection is the process by which a node determines that a collision has occurred.

- **LAN—local-area network**

A *local-area network* (LAN) is a computer network that spans a relatively small area. Most often, a LAN is confined to a single room, building or group of buildings, however, one LAN can be connected to other LANs over any distance via telephone lines and radio waves.

- **MAN—metropolitan area network**

A data network designed for a town or city. Geographically, MANs are larger than local-area networks (LANs), but smaller than wide-area networks (WANs).

- **WAN—wide area network**

A wide-area network (WAN) spans a relatively large geographical area and typically consists of two or more local-area networks (LANs).

- **Computer virus**

A computer virus is a program or piece of code that is loaded onto your computer without your knowledge and runs against your wishes.

● **Bandwidth**

Bandwidth is defined as the amount of data that can be transmitted in a fixed amount of time or range within a band of frequencies or wavelengths.

● **Twisted-pair cable**

A type of cable that consists of two independently insulated wires twisted around one another.

● **Fiber optics**

A technology that uses glass (or plastic) threads (fibers) to transmit data. Fiber optics has several advantages over traditional metal communications lines.

● **Packet**

A piece of a message transmitted over a packet-switching network.

● **Packet switching**

A protocol in which messages are divided into packets before they are sent. Each packet is then transmitted individually.

New Words

topology	[tə'pɔlədʒi]	n. 拓扑结构；拓扑（学）
endpoint	['endpɔit]	n. 端点，终点
incompatibility	[,inkəm,pætə'biləti]	n. 不兼容；不一致
ring	[riŋ]	n. 戒指，指环
loop	[lu:p]	n. 回路；圈；循环
routing	['ru:tiŋ]	vt. 按某路线发送，路由
source	[sɔ:s]	n. 根源，本源
destination	[,desti'neiʃn]	n. 目的，目的地，终点
dedicate	['dedikeitid]	adj. 专用的；专注的
optimal	['ɔptiməl]	adj. 最佳的，最优的
efficiency	[i'fiʃnsi]	n. 功效；效率
hub	[hʌb]	n. 轮轴；中心；集线器
forward	['fɔ:wəd]	vt. 转寄；发送
router	['ru:tə(r)]	n. 路由器
gateway	['geitwei]	n. 入口；网关
segment	['segmənt]	n. 环节；部分；网段
switch	[switʃ]	n. 开关；转换；交换机
repeater	[ri'pi:tə(r)]	n. 中继器
regenerate	[ri'dʒenəreit]	vt. & vi. 回收；使再生

replicate	[ˈreplikeit]	vt. 复制，复写
convention	[kənˈvenʃn]	n. 规矩；惯例，习俗
determine	[diˈtɜ:min]	vt. 决定，确定
mesh	[meʃ]	n. 网眼，网格
collision	[kəˈliʒn]	n. 碰撞；冲突
detection	[diˈtekʃn]	n. 侦查；检查
virus	[ˈvairəs]	n. 病毒；计算机病毒
bandwidth	[ˈbændwidθ]	n. 带宽
wavelength	[ˈweivleŋθ]	n. 波段；波长
frequency	[ˈfri:kwənsi]	n. 频率，次数
fiber	[ˈfaibə]	n. 光纤；纤维
optic	[ˈɔptik]	adj. 光学的；眼睛的
twisted	[ˈtwistid]	v. 扭，搓，缠绕
cable	[ˈkeibl]	n. 绳索；电缆
plastic	[ˈplæstik]	n. 整形；塑料制品
metal	[ˈmetl]	n. 金属；金属元素
individually	[ˌindiˈvidʒuəli]	adv. 分别地；各个地

Phrases

star topology	星形拓扑结构
bus network	总线网
packet protocol	分组协议
data link layer	数据链路层
application sharing	应用共享
network protocol	网络协议
network topology	网络拓扑结构
collision detection	冲突检测
computer virus	计算机病毒
twisted-pair cable	双绞线电缆
fiber optics	光纤
packet switching	分组交换

Abbreviation

LAN—Local Area Network   局域网

MAN—Metropolitan Area Network　　　城域网
WAN—Wide Area Network　　　广域网

Exercises

Ex1. Multiple choice.

1. Modulation and demodulation are the processes of a(n) (　　).
   A. connection device　　B. node　　C. modulator　　D. modem
2. Standard telephone lines and conventional modems provide what is called (　　).
   A. network architecture　　B. broadband　　C. dial-up service　　D. data transmission
3. In (　　), bits flow in a series or continuous stream.
   A. parallel data transmission　　　　B. packets
   C. simplex communication　　　　D. half-duplex communication
   E. serial data transmission
4. Every computer on the Internet has a unique numeric address called a(n) (　　).
   A. packet　　　　B. RS-232C connector　　　　C. IP address
   D. bandwidth　　E. network bridge
5. The standard protocol for the Internet is (　　).
   A. TCP/IP　　B. OSI　　C. RW-232C　　D. DSL
   E. NOS
6. (　　) describes how the network is arranged and how the resources are coordinated and shared.
   A. Topology　　　　　　　　B. Communication channel
   C. Sharing system　　　　　D. Network architecture
7. (　　) controls and coordinates the activities of all computers and devices on a network.
   A. TCP/IP　　B. NOS　　C. DNS　　D. OSI
   E. none of the above
8. In a (n) (　　) network, each device in the network handles its own communications control.
   A. host　　B. client　　C. bus　　D. sharing
   E. polling
9. Computer (　　) is a complex consisting of two or more connected computing units, it is used for the purpose of data communication and resource sharing.
   A. storage　　B. device　　C. network　　D. processor
10. One solution to major security problems is (　　), which are frequently installed to fix known security holes.
    A. patches　　B. compensations　　C. complements　　D. additions
11. All parties involved in a communication must agree on a set of rules to be used when __(1)__ messages, including the language to be used and the rules for when messages can

be sent. Diplomats called such an (2) a protocol. The term is applied to computer communication as well —a set of rules that (3) the format of messages and the appropriate actions required for each message is known as a network protocol or a computer communication protocol. The software that carries out such rules is (4) protocol software. An individual network protocol can be as simple as an agreement to use ASCII when transferring a text file, or as complex as an agreement to use a complicated mathematical (5) to encrypt data.

Words to be chosen from:
A. function   B. exchanging   C. agreement   D. called   E. specify
Answers: (1) _____ (2) _____ (3) _____ (4) _____ (5) _____

12. A network Interface Card (NIC) is a computer circuit board or card that is (1) in a computer so that it can be connected to a network. PCs and (2) on LAN typically contains a NIC specifically designed for the LAN transmission technology, such as (3) or Token Ring. NIC provide a (4), full-time connection to a network. Most home and portable computers connect to the Internet through a (5) connection. The modem provides the connection interface to the ISP.

A. dial-up   B. Ethernet   C. dedicated   D. installed   E. workstations
Answers: (1) _____ (2) _____ (3) _____ (4) _____ (5) _____

Ex2. Match each numbered item with the most closely related lettered item.
a. coaxial cable   b. demodulation   c. analog signals   d. bandwidth
e. parallel data transmission f. distributed processing g. Bluetooth   h. half-duplex
i. packets   j. client   k. star network   l. fiber optic

(   ) 1. The type of signals that telephones typically send and receive.
(   ) 2. High-frequency transmission cable with a single solid-copper core.
(   ) 3. Uses microwaves to transmit data over short distances up to 33 feet.
(   ) 4. The process of converting analog to digital signals.
(   ) 5. Cable that transmits data as pulses of light through tubes of glass.
(   ) 6. Bits per second transmission capability of a channel.
(   ) 7. Transmission in which bits flow through separate lines simultaneously.
(   ) 8. Broken-down parts of a message that are sent over the Internet.
(   ) 9. A node that requests and uses resources available from other nodes.
(   ) 10. System in which computing power is located and shared at different locations.
(   ) 11. Several computers linked to a central host that serves as hosts to smaller computers or devices.
(   ) 12. Mode of data flow wherein data flows in both directions but not simultaneously.

Ex3. Computer English test.
1. Communication protocols are (1) connection-oriented or connectionless, (2) whether the sender of a message needs to contact and maintain a dialog with the recipient or (3) send a message without any prior connect and with the hope that the recipient receives everything (4). These methods (5) the two ways that communication is implemented on networks.

(1) A. not   B. neither   C. either   D. all

(2) A. fulled   B. flooded   C. depending on   D. defined by
(3) A. Immediately   B. simply   C. accordingly   D. properly
(4) A. in order   B. in array   C. in series   D. in queueing
(5) A. make known   B. disclose   C. reveal   D. discover

Answers: (1) _____ (2) _____ (3) _____ (4) _____ (5) _____

2. Heterogeneous network environments consist of computer systems from __(1)__ vendors that run __(2)__ operating systems and communication protocols. An organization that __(3)__ its computer resources is usually __(4)__ the task of integrating its heterogeneous systems. Typically, each department or division has defined its own network needs __(5)__ OS, LAN topology, communication protocols, applications, and other components.

(1) A. same   B. similar   C. different   D. difference
(2) A. same   B. similar   C. different   D. difference
(3) A. consolidate   B. consists   C. considerate   D. consoles
(4) A. faced on   B. faced with   C. faced about   D. faced up to
(5) A. in general   B. in any term   C. in set terms   D. in terms of

Answers: (1) _____ (2) _____ (3) _____ (4) _____ (5) _____

Ex4. Translate the following sentences into English.

1. 服务器与其他节点共享资源，取决于共享的资源，它可以被称为文件服务器、打印服务器、通信服务器、Web 服务器或数据库服务器。

2. 物理连接使用固体介质，如双绞电话线、同轴电缆、光纤电缆。无线连接使用空气而不是固体物质来连接发送与接收设备，两种主要的无线技术是微波和卫星。

3. 在总线网中，网络中的每个设备都会处理自己的通信控制。没有主机，所有的通信沿着被称为总线的公共连接电缆传送。当信息经过总线时，每个设备都会检查该信息是否是给它的。

Ex5. Speaking.

1. What is computer network?
2. What are the benefits of connecting computers and peripherals in a network?
3. How can a PC be connected to another computer?
4. Identify and describe the various physical and wireless communication channels.
5. Define and discuss the four principle network topologies.
6. What do the terms means in your language: LAN, MAN, WAN?
7. Which ways can a virus enter a computer system?

Part II   Reading materials

Network security

Network security consists of the policies and practices adopted to prevent and monitor unauthorized access, misuse, modification, or denial of a computer network and network-accessible resources. Network security involves the authorization of access to data in a network, which is controlled by the network administrator. Users choose or are assigned an ID and password or other authenticating information that allows them access to information and programs within their authority. Network security covers a variety of computer networks, both public and private, that are used in everyday jobs; conducting transactions and communications among businesses, government agencies and individuals. Networks can be private, such as within a company, and others which might be open to public access. Network security is involved in organizations, enterprises, and other types of institutions. It does as its title explains: It secures the network, as well as protecting and overseeing operations being done. The most common and simple way of protecting a network resource is by assigning it a unique name and a corresponding password.

Network security starts with authenticating, commonly with a username and a password. Since this requires just one detail authenticating the user name—i.e., the password—this is sometimes termed one-factor authentication. With two-factor authentication, something the user "has" is also used (e.g., a security tokenor "dongle", an ATM card, or a mobile phone); and with three-factor authentication, something the user "is" is also used (e.g., a fingerprint or retinal scan).

Once authenticated, a firewall enforces access policies such as what services are allowed to be accessed by the network users. Though effective to prevent unauthorized access, this component may fail to check potentially harmful content such as computer worms or Trojans being transmitted over the network. Anti-virus software or anintrusion prevention system (IPS) help detect and inhibit the action of such malware. An anomaly-based intrusion detection system may also monitor the network like wireshark traffic and may be logged for audit purposes and for later high-level analysis. Newer systems combining unsupervised machine learning with full network traffic analysis can detect active network attackers from malicious insiders or targeted external attackers that have compromised a user machine or account.

Communication between two hosts using a network may be encrypted to maintain privacy.

术语参考译文

网络
网络是两个或多个连接在一起的计算机系统集合。网络上的计算机有时被称为节点。为网络分配资源的计算机和设备被称为服务器。

星形网
在使用星形拓扑结构的局域网（LAN）中，所有节点都连接到一台中央计算机。

环网
一种拓扑结构为环形的局域网。即所有的节点连接在一个封闭的环上。

总线网
一种网络，其中所有节点连接到一个有两个端点的单线（总线）。例如，以太网 10base-2

和 10base-5 是总线网。

网络标准

网络标准通过定义网络设备之间的通信规则来确保网络技术的互操作性。网络标准的存在有助于确保不同厂商的产品能够在网络中协同工作，而不存在不兼容的风险。

路由

路由是指将网络数据包从其源移动到目的地的过程。路由通常由一个称为路由器的专用网络设备完成，是因特网的一个重要特征，它使消息从一台计算机传递到另一台计算机，最终到达目标机。

通过在网络间转发网络流量，并为在网络上传送的数据选择最佳的路径，路由确保最佳的网络效率。

集线器

网络中设备的公共连接点。集线器作为所有网络设备的中心连接，并处理称为帧的数据类型。

路由器

一种沿网络转发数据包的设备。路由器至少连接两个网络并位于网关上。

交换机

在网络中，在局域网段之间过滤和转发数据包的设备。交换机工作在数据链路层（第2层），有时在 OSI 参考模型的网络层（第3层），因此支持任何分组协议。

中继器

该网络设备用于再生或复制传输系统中使用的信号，以再生由于传输损耗而失真的信号。

应用共享

许多视频会议应用的一项功能，使与会者能够同时运行同一应用程序。

网络协议

网络协议是规则和约定的语言，它被用来处理网络设备之间的通信并确保网络的最佳运行。网络协议包括关键的 Internet 协议，如 IP、IPv6 以及 DNS、FTP，它还包括更多特定的网络协议，如 SNMP 和 NTP。

网络拓扑

网络拓扑结构是指网络上连接设备的布局。网络中不同节点怎样相互连接以及它们如何通信是由网络拓扑结构决定的。网络拓扑可以应用于物理或逻辑布局，常见的网络拓扑结构包括总线、环、树、星和网格。

冲突检测

冲突检测是一个节点确定冲突发生的过程。

局域网

局域网（LAN）是一个跨越相对较小区域的计算机网络。大多数情况下，局域网仅限于一个房间、建筑物或一组建筑物，但是，通过电话线和无线电波，一个局域网可以连接到任何地方的其他局域网。

城域网

城域网（MAN）是一个为城镇或城市设计的数据网络。从地理位置上说，城域网比局域网（LAN）大，但比广域网（WAN）小。

广域网

广域网（WAN）跨越一个相对较大的地理区域，通常由两个或多个局域网（LAN）组成。

计算机病毒（病毒）

计算机病毒是一个程序或一块代码，在你不知情的情况下被加载到计算机上，违背你的意愿运行。

带宽

带宽是指在一定数量的时间或者在一定频率或波长范围内传输的数据量。

双绞线电缆

一种由两个独立绝缘电线相互缠绕的电缆。

光学纤维

用玻璃（或塑料）线（纤维）传输数据的技术。光纤比传统的金属通信线优点多一些。

分组

在分组交换网络上传输的一条消息。

分组交换

在发送之前将报文分成包的协议，然后每个数据包单独发送。

Unit 9　Internet and Online Services

Part I　Computer terms

- **Internet**

The Internet is a global networking infrastructure that connects millions of computers together, forming a network in which any computer can communicate with any other computer as long as they are both connected to the Internet.

- **Intranet**

An intranet is a network based on TCP/IP protocols belonging to an organization, accessible only by the organization's members or those with authorization.

- **Chat**

Online chat refers to real-time communication that occurs between two users over the computer. There are specific chat programs, like IRC, instant messaging, Facebook chat or even email that allows people to participate in private or group text-based conversations.

- **Instant messaging**

A type of communications service used to create a private chat room with another individual in order to communicate in real time over the Internet.

- **Domain name**

Domain names are used to identify one or more IP addresses. For example, the domain name microsoft.com represents about a dozen IP addresses. Domain names are used in URLs to identify particular Web pages. For example, in the URL *http://www.pcwebopedia.com/index. html,* the domain name is *pcwebopedia.com.*

Every domain name has a suffix that indicates which top level domain (TLD) it belongs to. There are only a limited number of such domains. For example:

- gov—Government agencies
- edu—Educational institutions
- org—Organizations (nonprofit)
- mil—Military
- com—commercial business
- net—Network organizations
- ca—Canada
- th—Thailand

Because the Internet is based on IP addresses, not domain names, every Web server requires a Domain Name System (DNS) server to translate domain names into IP addresses.

- **IP address—Internet Protocol (IP) address**

An IP address is an identifier for devices on a TCP/IP network. Networks using TCP/IP route messages based on the IP address of the destination.

- **Electronic commerce**

Often referred to as ecommerce, the phrase describes business that is conducted over the Internet using any of the applications that rely on the Internet, including email, shopping carts, social media and Web services, among others.

- **Electronic mail**

Electronic mail (also called email or e-mail) is the transmission of messages over communications networks. Most email systems include an editor for composing messages and then you send the message by specifying the recipient's email address. Although different email systems use different formats, there are standards that make it possible for users on all systems to exchange messages.

- **Attachment**

An attachment is a file attached to an email message. Many email systems support sending text files as email.

- **Search engines**

Web search engines work by sending out a spider to fetch as many documents as possible. Another program, called an indexer, then reads these documents and creates an index based on the words contained in each document. Each search engine — like Google, Bing, Ask.com and others — use proprietary algorithms to crawl, index and display search results.

- **Web browser (browser)**

Web browser is a software application used to locate, retrieve and display content on the World Wide Web, including Web pages, images and video.

- **FTP—file transfer protocol**

File Transfer Protocol (FTP) is the commonly used protocol for exchanging files over the Internet. FTP works in the same way as HTTP and SMTP.

- **Web server**

Web servers are computers that deliver (serves up) Web pages. Every Web server has an IP address and possibly a domain name.

- **HTTP—HyperText Transfer Protocol**

Short for HyperText Transfer Protocol, HTTP is the underlying protocol used by the World Wide Web to define how messages are formatted and transmitted.

- **ISP—Internet service provider**

Short for Internet Service Provider, it refers to a company that provides Internet services, including personal and business access to the Internet.

- **TCP/IP—Transmission Control Protocol/Internet Protocol**

TCP/IP is the suite of communications protocols used to connect hosts on the Internet. TCP/IP uses several protocols, the two main ones being TCP and IP. TCP/IP is built into the

UNIX operating system and is used by the Internet, making it the defacto standard for transmitting data over networks. Even network operating systems that have their own protocols, such as Netware, also support TCP/IP.

New Words

infrastructure	['infrəstrʌktʃə(r)]	n. 基础设施；基础建设
intranet	['intrənet]	n. 内联网
authorization	[ˌɔːθəraɪ'zeɪʃn]	n. 授权，批准
chat	[tʃæt]	vi. 聊天；闲谈
conversation	[ˌkɔnvə'seɪʃn]	n. 交谈，会话
private	['praɪvət]	adj. 私有的，秘密的
individual	[ˌindi'vidʒuəl]	adj. 个人的；个别的
domain	[də'meɪn]	n. 范围，领域；域名
web page	[web peɪdʒ]	n. 网页
nonprofit	[ˌnɔn'prɔfit]	adj. 非营利的
institution	[ˌinsti'tjuːʃn]	n. 机构；制度
agency	['eɪdʒənsi]	n. 代理；机构
phrase	[freɪz]	n. 成语；说法；短语
cart	[kɑːt]	n. 运货马车，手推车
attachment	[ə'tætʃmənt]	n. 附件，附属物
format	['fɔːmæt]	n. 版式；形式
index	['indeks]	n. 索引；[数]指数；指示
browser	['braʊzə(r)]	n. 浏览器

Phrases

educational institution	教育机构
government agency	政府机构
commercial business	商业企业
network organization	网络组织
electronic commerce	电子商务
electronic mail	电子邮件
shopping cart	购物车
search engine	搜索引擎
Web browser	Web 浏览器
Web server	Web 服务器

Abbreviation

TCP/IP—Transfer Control Protocol/Internet Protocol　　传输控制协议/网际协议
FTP—File Transfer Protocol　　文件传输协议
HTTP—HyperText Transfer Protocol　　超文本传输协议
ISP—Internet Service Provider　　因特网服务提供商

Exercises

Ex1. Multiple choice.

1. The Internet was launched in 1969 when the U.S. funded a project to develop a national computer network called (　　).

　　A. APARNET　　B. CERN　　C. WWW　　D. the Web

2. In (　　) commerce, individuals sell to other individuals without ever meeting face to face.

　　A. C2C　　B. B2C　　C. B2B　　D. C2I

3. (　　) is a clickable string or graphic that points to another Web page or document.

　　A. Link　　B. Anchor　　C. Browser　　D. Hyperlink

4. One of the greatest features of a home (　　) is the ability to share one Internet connection simultaneously over two or more computers.

　　A. computer　　B. device　　C. network　　D. work

5. One of the basic rules of computer security if to change your (　　) regularly.

　　A. name　　B. computer　　C. device　　D. password

6. The (　　) in E-mail messages has affected almost every computer around the world and has caused the damage of up to US$ 1 billion in North America.

　　A. illness　　B. virus　　C. weakness　　D. attachment

7. The (　　) is the collection of computers connected together by phone lines that allows for the global sharing of information.

　　A. interface　　B. Internet　　C. LAN　　D. WWW

8. A firewall is a __(1)__ system designed to __(2)__ an organization's network against threats.

　　(1) A. operating　　B. programming　　C. security　　D. service
　　(2) A. prevent　　B. protect　　C. develop　　D. exploit

9. Which of these statements about connecting to the Internet is true?（　　）

　　A. The Internet can only be used to link computers with same operating system.
　　B. The Internet can be used to connect computers with different ISP
　　C. You must have a modem to connect to the Internet
　　D. You must have a telephone line to connect to the Internet

10. (   ) is the sending and receiving of the messages by computer. It is a fast, low-cost way of communicating worldwide.

    A. LAN   B. Post Office   C. E-mail   D. Interface

11. Each electronic mailbox has a __(1)__ electronic mail address, which is usually called an E-mail address. When someone sends a message, he uses an E-mail address to __(2)__ the recipient. Each E-mail address represents both a mailbox and a host computer. One widely used __(3)__ is the form *mailbox@computer*, where *mailbox* is a string that denotes a user's mailbox, and *computer* is a string that __(4)__ to the computer on which the mailbox is __(5)__.

Words to be chosen from:

    A. located   B. format   C. unique   D. specify   E. refers

Answers: (1)_____ (2)_____ (3)_____ (4)_____ (5)_____

12. The software that translates computer domain names into __(1)__ Internet addresses provided an interesting example of client-server interaction. The DB of names is not kept on a single computer. Instead, the naming information is __(2)__ among a potentially large set of servers located at sites __(3)__ the Internet. Whenever an application program needs to translate a name, the application becomes a __(4)__ of the naming system. The client sends a request message to a name server, which finds the corresponding address and sends a reply message. If it cannot answer a request, a name server __(5)__ becomes the client of another name server, until a server if found that can answer the request.

Words to be chosen from:

    A. client   B. temporarily   C. kept   D. across   E. equivalent

Answers: (1)_____ (2)_____ (3)_____ (4)_____ (5)_____

Ex2. Match each numbered item with the most closely related lettered item.

    a. surfing   b. header   c. Web page   d. ISP

    e. FTP   f. hypermedia   g. IP address   h. URLs

    i. browser   j. search engine   k. signature line l. e-commerce

(   ) 1. An organization that provides services for accessing, using, or participating in the Internet.

(   ) 2. Addresses of Web resources.

(   ) 3. Moving from one Web site to another.

(   ) 4. A software application for retrieving, presenting and traversing information resources on the World Wide Web.

(   ) 5. Part of an e-mail message that includes the subject, address, and attachments.

(   ) 6. Typically includes the sender's name, address, and telephone number.

(   ) 7. A standard network protocol used for the transfer of computer filesfrom a server to a client using the Client–server model on a computer network.

(   ) 8. Buying and selling goods over the Internet.

(   ) 9. A numerical label assigned to each device (e.g., computer, printer) participating in a computer network that uses the Internet Protocol for communication.

(   ) 10. An extension of the term hypertext, is a nonlinear medium of information which includes graphics, audio, video, plain text and hyperlinks.

(   ) 11. A document that is suitable for the World Wide Web and web browsers.

(   ) 12. A software system that is designed to search for information on the World Wide Web.

Ex3. Computer English test.

1. The major problem with E-mail is that it is __(1)__ easy to use that people can become __(2)__ with messages __(3)__ they can possibly answer in a day. In addition, mail boxes require some management to __(4)__ messages or archive those that might be required later. Senders don't always know about your E-mail backlog and often send __(5)__ messages.

(1) A. too   B. so   C. very   D. much
(2) A. full   B. lost   C. inundated   D. filled
(3) A. more than   B. than   C. that   D. which
(4) A. manage   B. save   C. backup   D. dispose of
(5) A. too many   B. redundant   C. long   D. trivial

Answers: (1)_____ (2)_____ (3)_____ (4)_____ (5)_____

2. Communication via e-mail is by far the most common Internet activity. You can __(1)__ with anyone in the world who has an Internet address or e-mail __(2)__ with a system connection to the Internet. All you need is __(3)__ the Internet and an e-mail program. Two of the most widely used e-mail programs are Microsoft's __(4)__ and Netscape's Navigator. A typical E-mail message has three basic __(5)__: header, message, and signature.

(1) A. connect   B. exchange   C. communicate   D. game
(2) A. account   B. No.   C. user   D. administrator
(3) A. attach with   B. grasp   C. possess   D. access to
(4) A. Yahoo   B. Explorer   C. Foxmail   D. Outlook Express
(5) A. packets   B. elements   C. frames   D. cells

Answers: (1)_____ (2)_____ (3)_____ (4)_____ (5)_____

Ex4. Translate the following sentences into English.

1. 软盘是一种便携式的存储介质，它们被用于存储和传输字处理、电子表格和其他类型的文件。

2. 文件压缩和解压缩通过减少存储数据和程序所需空间的数量来提高存储容量。文件压缩不局限于硬盘系统，它也常用于压缩软盘上的文件。

3. 闪存是固态存储器的一个例子，它通常用于存储数字化图像和记录 MP3 文件。

4. 磁带提供顺序访问，磁盘提供直接访问，磁带主要用于备份数据。

Ex5. Speaking.

1. Discuss the uses of Internet. Which activities have you participated in? Which one do you

think is the most popular?

2. What are the basic elements of and e-mail message?

3. What are the principle measures used to protect computer security? What is encryption?

4. What is ecommerce? Which activities relate to ecommerce have you participated in? How do you pay on Internet?

5. Why security is so important on the Internet?

6. What do the terms means in your language: IP address, domain name, Web browser, HTTP?

Part II   Reading materials

Internet

The Internet is the global system of interconnected computer networks that use the Internet protocol suite(TCP/IP) to link devices worldwide. It is a network of networks that consists of private, public, academic, business, and government networks of local to global scope, linked by a broad array of electronic, wireless, and optical networking technologies. The Internet carries an extensive range of information resources and services, such as the inter-linked hypertext documents and applications of the World Wide Web (WWW), electronic mail, telephony, and peer-to-peer networks for file sharing.

The origins of the Internet date back to research commissioned by the United States federal government in the 1960s to build robust, fault-tolerant communication via computer networks. The primary precursor network, the ARPANET, initially served as a backbone for interconnection of regional academic and military networks in the 1980s. The funding of the National Science Foundation Network as a new backbone in the 1980s, as well as private funding for other commercial extensions, led to worldwide participation in the development of new networking technologies, and the merger of many networks. The linking of commercial networks and enterprises by the early 1990s marks the beginning of the transition to the modern Internet, and generated a sustained exponential growth as generations of institutional, personal, and mobile computers were connected to the network. Although the Internet was widely used by academia since the 1980s, the commercialization incorporated its services and technologies into virtually every aspect of modern life.

Internet use grew rapidly in the West from the mid-1990s and from the late 1990s in the developing world. In the 20 years since 1995, Internet use has grown 100-times, measured for the period of one year, to over one third of the world population. Most traditional communications media, including telephony, radio, television, paper mail and newspapers are being reshaped or redefined by the Internet, giving birth to new services such as email, Internet telephony, Internet television music, digital newspapers, and video streamingwebsites. Newspaper, book, and other print publishing are adapting to website technology, or are reshaped into blogging, web feeds and online news aggregators. The entertainment industry was initially the fastest growing segment on the Internet. The Internet has enabled and accelerated new forms of personal interactions through instant messaging, Internet forums, and social networking. Online shopping has grown exponentially both for major retailers and small businesses and entrepreneurs, as it enables firms to extend their "bricks and mortar" presence to serve a larger market or even sell goods and services entirely online.Business-to-business and financial services on the Internet affect supply chains across entire industries.

The Internet has no centralized governance in either technological implementation or policies for access and usage; each constituent network sets its own policies. Only the overreaching definitions of the two principal name spaces in the Internet, the Internet Protocol address space and the Domain Name System (DNS), are directed by a maintainer organization, the Internet Corporation for Assigned Names and Numbers (ICANN). The technical underpinning and standardization of the core protocols is an activity of the Internet Engineering Task Force(IETF), a non-profit organization of loosely affiliated international participants that anyone may associate with by contributing technical expertise.

eCommerce Resources

E-commerce is a transaction of buying or selling online. Electronic commerce draws on technologies such as mobile commerce,electronic funds transfer,supply chain management, Internet marketing, online transaction processing,electronic data interchange (EDI), inventory management systems, and automated data collectionsystems. Modern electronic commerce typically uses the World Wide Web for at least one part of the transaction's life cycle although it may also use other technologies such as e-mail.

E-commerce businesses may employ some or all of the following:
- Online shopping web sites for retail sales direct to consumers
- Providing or participating in online marketplaces, which process third-party business-to-consumer or consumer-to-consumer sales
- Business-to-business buying and selling
- Gathering and using demographic data through web contacts and social media
- Business-to-business (B2B) electronic data interchange
- Marketing to prospective and established customers by e-mail or fax (for example, with newsletters)
- Engaging in pretail for launching new products and services
- Online financial exchanges for currency exchanges or trading purposes

Some common applications related to electronic commerce are:
- Document automation in supply chain and logistics
- Domestic and international payment systems
- Enterprise content management
- Group buying

- Print on demand
- Automated online assistant
- Newsgroups
- Online shopping and order tracking
- Online banking
- Online office suites
- Shopping cart software
- Teleconferencing
- Electronic tickets
- Social networking
- Instant messaging
- Pretail
- Digital Wallet

术语参考译文

因特网

因特网是一个全球性的网络基础设施,将数以百万计的计算机连接在一起,形成一个网络,其中任何计算机都可以与其他计算机通信,只要它们都连接到因特网。

内联网

内联网是一个属于组织、基于 TCP／IP 协议的网络,只有组织的成员或授权才可以访问。

聊天

在线聊天是指两个用户之间通过计算机进行实时通信。有一些专门的聊天程序,如 IRC、即时通讯、Facebook 聊天甚至电子邮件,让人们参加基于文本的私人或团体对话。

即时通讯

一种用于与另一个人建立私人聊天室以便在因特网上实时通信的通信服务。

域名

域名用于识别一个或多个 IP 地址。例如,域名 microsoft.com 代表了大约十几个 IP 地址。在 URL 中,域名用于识别特定的网页。例如,在 url http://www.pcwebopedia.com/index.html,域名为 pcwebopedia.com。

每个域名后缀表示它属于哪个顶级域名(TLD)。只有有限数量的这样的域。例如,

- gov—政府机构
- edu—教育机构
- org—组织(非营利)
- mil—军事
- com—商业业务
- net—网络组织

- ca—加拿大
- th—泰国

由于互联网基于 IP 地址，而不是域名，每个 Web 服务器都需要一个域名系统（DNS）服务器将域名翻译成 IP 地址。

IP 地址——网际协议（IP）地址

IP 地址是一个用于 TCP/IP 网络设备的标识符。基于目的地 IP 地址，网络使用 TCP/IP 转发信息。

电子商务

eCommerce 通常被称为电子商务，这句话描述的业务是在因特网上使用任何依赖于因特网的应用程序，包括电子邮件、购物车、社交媒体和 Web 服务，等等。

电子邮箱

电子邮件是通过通信网络传输信息。大多数电子邮件系统包括编辑邮件，然后通过指定收件人的电子邮件地址发送邮件。虽然不同的电子邮件系统使用不同的格式，但有标准使得所有系统的用户都可以交换信息。

附件

附件是附在电子邮件上的文件。许多电子邮件系统都支持用电子邮件发送文本文件。

搜索引擎

网络搜索引擎通过发送一个"蜘蛛"来获取尽可能多的文档。然后另一个称为索引器的程序，读取这些文件并基于每个文件中所含关键词创建一个索引。每个搜索引擎如谷歌、Bing、Ask.com 等，使用专有的算法来抓取、索引和显示搜索结果。

Web 浏览器（浏览器）

Web 浏览器是一个软件应用程序，用于在万维网定位、检索和显示内容，包括网页、图像和视频。

文件传输协议

文件传输协议（FTP）是因特网上用于交换文件的常用协议。FTP 与 HTTP、SMTP 工作方式相同。

Web 服务器

网络服务器是传送（服务）网页的计算机。每个 Web 服务器都有一个 IP 地址和一个域名。

超文本传输协议

超文本传输协议（HTTP）是万维网用来定义信息如何被格式化和传输的基本协议。

互联网服务提供商

ISP 是互联网服务提供商的简称，它指的是一家提供互联网服务的公司，包括个人和企业上网。

TCP/IP——传输控制协议/网际协议

TCP/IP 是用于连接因特网上主机的一套通信协议。TCP/IP 使用多个协议，最重要的两个是 TCP 和 IP。TCP/IP 协议内置于 UNIX 操作系统，并在因特网上使用，成为在网络上传输数据事实上的标准。即使网络操作系统有自己的协议，如 Netware，它也支持 TCP/IP。

Unit 10　World Wide Web

Part I　Computer terms

- **Web/World Wide Web**

The Web, or World Wide Web, is basically a system of Internet servers that support specially formatted documents. The documents are formatted in a markup language called HTML (*HyperText Markup Language*) that supports links to other documents, as well as graphics, audio, and video files.

This means you can jump from one document to another simply by clicking on hot spots. Not all Internet servers are part of the World Wide Web.

- **Internet address**

An Internet address uniquely identifies a node on the Internet. Internet address may also refer to the name or IP of a Web site (URL).

- **Hyperlink**

An element in an electronic document that links to another place in the same document or to an entirely different document.

- **Hypermedia**

An extension to hypertext that supports linking graphics, sound, and video elements in addition to text elements.

- **Web page**

A document on the World Wide Web. Every Web page is identified by a unique URL (Uniform Resource Locator).

- **Java applet**

An applet is a small Internet-based program written in Java, a programming language for the Web, which can be downloaded by any computer.

- **JavaScript**

A scripting language developed by Netscape to enable Web authors to design interactive sites.

- **CSS—Cascading Style Sheets**

Short for Cascading Style Sheets, a new feature being added to HTML that gives both Web site developers and users more control over how pages are displayed. With CSS, designers and users can create style sheets that define how different elements, such as headers and links, appear. These style sheets can then be applied to any Web page.

The term cascading derives from the fact that multiple style sheets can be applied to the same Web page. CSS was developed by the W3C.

- **Website**

A site (location) on the World Wide Web. Each website contains a home page, which is the first document users see when they enter the site.

- **Domain name**

Domain names are used to identify one or more IP addresses. For example, the domain name microsoft.com represents about a dozen IP addresses.

- **DNS—Domain Name System**

Short for Domain Name System (or Service or Server), an Internet service that translates domain names into IP addresses.

- **Host**

A computer system that is accessed by a user working at a remote location. Typically, the term is used when there are two computer systems connected by modems and telephone lines.

- **URL—Uniform Resource Locator**

Abbreviation of Uniform Resource Locator (URL) it is the global address of documents and other resources on the World Wide Web. For example, www.webopedia.com is a URL. A URL is one type of Uniform Resource Identifier (URI).

The first part of the URL is called a *protocol identifier* and it indicates what protocol to use, and the second part is called are *source name* and it specifies the IP address or the domain name where the resource is located. The protocol identifier and the resource name are separated by a colon and two forward slashes.

- **Home page**

The main page of a Web site. Typically, the home page serves as an index or table of contents to other documents stored at the site.

- **Frames**

A feature supported by most modern Web browsers than enables the Web author to divide the browser display area into two or more sections (frames). The contents of each frame are taken from a different Web page.

- **Tag**

A command inserted in a document that specifies how the document, or a portion of the document, should be formatted.

- **Web authoring**

A category of software that enables the user to develop a Web site in a desktop publishing format.

- **Container**

In HTML, the container is the area enclosed by the beginning and ending tags. For example < HTML >encloses an entire document while other tags may enclose a single word, paragraph, or other elements.

- **Mouseover**

A JavaScript element that triggers a change on an item (typically a graphic change, such as making an image or hyperlink appear) in a Web page when the pointer passes over it. The change usually signifies that the item is a link to related or additional information.

- **Static**

Generally refers to elements of the Internet or computer programming that are fixed and not capable of action or change. The opposite of static is dynamic.

- **Dynamic HTML**

Refers to Web content that changes each time it is viewed. For example, the same URL could result in a different page depending on any number of parameters, such as:

(1) Geographic location of the reader

(2) Time of day

(3) Previous pages viewed by the reader

(4) Profile of the reader

There are many technologies for producing dynamic HTML, including CGI scripts, Server-Side Includes (SSI),cookies, Java, JavaScript, and ActiveX.

New Words

spot	[spɔt]	n. 地点，场所
hypermedia	[ˌhaipə'mi:diə]	n. 超媒体
extension	[ik'stenʃn]	n. 伸展，扩大
hyperlink	['haipəliŋk]	n. 超链接
applet	['æplət]	n. Java 小应用程序
download	[ˌdaun'ləud]	v. 下载
website	['websait]	n. 网站
host	[həust]	n. 主机；主人，东道主
home page	[həum peidʒ]	n. 主页，首页
container	[kən'teinə(r)]	n. 容器；箱
enclose	[in'kləuz]	vt. 把……围起来；把……装入信封
mouseover	[maus 'əuvə(r)]	n. 悬停；鼠标经过
signify	['signifai]	vt. 意味，预示
static	['stætik]	adj. 静止的，静态的
dynamic	[dai'næmik]	adj. 动态的；动力的，动力学的
profile	['prəufail]	n. 侧面，轮廓；人物简介
geographic	[ˌdʒi:ə'græfik]	adj. 地理学的，地理的

Phrases

Internet address	因特网地址
Java applet	Java 小程序
home page	首页
Web authoring	Web 创作
dynamic HTML	动态超文本标记语言
geographic location	地理位置

Abbreviation

URL—Uniform Resource Locator　　　　统一资源定位器
URI—Uniform Resource Identifier　　　　统一资源标识符
CSS—Cascading Style Sheets　　　　　　层叠样式表
DNS—Domain Name System　　　　　　域名系统

Exercises

Ex1. Multiple choice.

1. (　) is the conscious effort to make all jobs similar, routine, and interchangeable.
   A. WWW　　　B. Information　　　C. Computerization　　　D. Standardization

2. A Web (　) is one of many software applications that function as the interface between a user and the Internet.
   A. display　　　B. browser　　　C. window　　　D. view

3. (　) are Web sites that search the Web for occurrences of a specified word or phrase.
   A. Search engine　　B. WWW　　　C. Internet　　　D. Java

4. (　) are programs that are automatically loaded and operate as a part of your browser.
   A. Add-ins　　　B. Plug-ins　　　C. Helpers　　　D. Utilities

5. （　）are those programs that help find the information you are trying to locate on the WWW.
   A. Windows　　　B. Search Engines　　　C. Web Sites　　　D. Web Pages

6. (　) means "Any HTML document on an HTTP Server".
   A. Web Server　　　B. Web page　　　C. Web Browser　　　D. Web site

7. Technically, the WWW is a __(1)__ hypertext system that uses the Internet __(2)__ its transport mechanism. In a hypertext system, you __(3)__ by clicking hyperlinks, which display another document which also contains __(4)__. What makes the Web such an exciting and useful medium is that the next document you see could be housed on computer next door or half-way

around the world. The Web makes the Internet easy to use. Created in 1989 at a research institute in Switzerland, the Web relies upon the hypertext transport protocol (HTTP), an Internet standard that __(5)__ how an application can locate and acquire resources stored on another computer on the Internet.

Words to be chosen from:

    A. specifies   B. global   C. hypertext   D. as   E. navigate

Answers: (1)_____ (2)_____ (3)_____ (4)_____ (5)_____

8. One of the main features of the Domain Name System is autonomy—the system is __(1)__ to allow each organization to computers or to change those names without informing a central authority. The naming hierarchy helps __(2)__ autonomy by allowing an organization to control all names with a particular suffix. Thus, Harvard University is free to create or change names with the end harvad.edu, __(3)__ Intel corporation is free to create or change names with intel.com. In addition to hierarchical names, the DNS uses client-server interaction to aid autonomy. In essence, the entire naming system operates as a large, distributed DB. Most organizations that have an Internet __(4)__ run a domain name server. Each server contains information that links the server to other domain name servers; the __(5)__ set of servers functions as a large, coordinated DB of names.

Words to be chosen from:

    A. resulting   B. connection   C. designed   D. achieve   E. while

Answers: (1)_____ (2)_____ (3)_____ (4)_____ (5)_____

9. TCP/IP stands __(1)__ Transmission Control Protocol/Internet Protocol. It was __(2)__ by the US Department of Defense as a method of transferring data between many different networks. IP, which the Internet is based __(3)__, is the protocol by which data is packaged and sent over the Internet. IP forwards each packet of information based on a 4-byte destination address or IP number. Packets are sent to gateway machines, which __(4)__ them according to their addresses. TCP verifies the transmission of data by the Internet Protocol. It is __(5)__ of detecting errors and lost data and of signaling re-transmission of flawed packets.

Words to be chosen from:

    A. capable   B. for   C. upon   D. developed   E. route

Answers: (1)_____ (2)_____ (3)_____ (4)_____ (5)_____

Ex2. Match each numbered item with the most closely related lettered item.

a. HTML   b. Trojan horse   c. rounter   d. hacker   e. protocol

f. firewall   g. worm   h. WWW   i. download   j. URL

(   ) 1. A networking device that forwards data packets between computer networks.

(   ) 2. A standard procedure for regulating data transmission between computers.

(   ) 3. To receive data from a remote system, typically a server such as a web server, an FTP server, an email server, or other similar systems.

(   ) 4. The standard markup language for creating web pages and web applications.

(   ) 5. Persons who gains unauthorized access to a computer system for the fun and challenge of it.

(　　) 6. A standalone malware computer program that replicates itself in order to spread to other computers.

(　　) 7. An information space where documents and other web resources are identified by Uniform Resource Locators (URLs), interlinked by hypertext links, and can be accessed via the Internet.

(　　) 8. Any malicious computer program which is used to hack into a computer by misleading users of its true intent.

(　　) 9. A reference to a web resource that specifies its location on a computer network and a mechanism for retrieving it.

(　　) 10. Security hardware and software that controls access to internal computer networks.

Ex3. Computer English test.

1. Because Web servers are platform and application __(1)__, they can send or request data from legacy or external applications including databases. All replies, once converted into __(2)__ mark-up language, can then be transmitted to a __(3)__. Used in this way, Intranets can __(4)__ lower desktop support costs, easy links with legacy applications and databases and, __(5)__ all, ease of use.

(1) A. coupled　　B. dependent　　C. independent　　D. related
(2) A. ciphertext　B. hypertext　　C. plaintext　　　D. supertext
(3) A. browser　　B. repeater　　　C. router　　　　D. server
(4) A. off　　　　B. offer　　　　　C. office　　　　D. officer
(5) A. abort　　　B. about　　　　　C. above　　　　D. around
Answers: (1)_____ (2)_____ (3)_____ (4)_____ (5)_____

2. An Intranet is __(1)__ the application of Internet technology within an internal or closed user group. Intranets are company __(2)__ and do not have to have a __(3)__ connection to the Internet. Used properly an Intranet is a highly effective corporate tool, capable of regularly __(4)__ information to empower the workforce with the information needed to perform their roles. Used in this way, Intranet represent a step towards the __(5)__ office.

（1）A. simple　　　B. simply　　　　C. single　　　　D. all
（2）A. common　　B. shared　　　　C. special　　　D. specific
（3）A. physical　　B. psychological　C. virtual　　　D. space
（4）A. update　　　B. updated　　　　C. updates　　　D. updating
（5）A. paper　　　B. paperbacked　　C. paperless　　D. paperwork
Answers: (1)_____ (2)_____ (3)_____ (4)_____ (5)_____

Ex4. Translate the following sentences into English.

1. 插件被自动加载并作为浏览器的一部分运行，一些插件包含在今天的许多浏览器中，另一些则必须安装。

2. 内联网是类似于因特网单位内部的私有网，它们使用仅对单位内部的人可用的浏览器和网页。

3. 防火墙是一个防止外部威胁的安全系统。它由硬件和软件构成。所有进出一个单位的通信都要通过一个称为代理服务器的专门的安全计算机。

Ex5. Speaking.

1. What is World Wide Web?
2. What do the terms means in your language: webpage, website, homepage, static webpage?
3. What is URL? Which parts do URL consist of?
4. What are Cascading Style Sheets? What are the advantages of CSS in Web Design?
5. What is dynamic HTML? Which technologies can be use to develop dynamic webpage?

Part II Reading materials

Understand Web applications

- **About web applications**

A web application is a website that contains pages with partly or entirely undetermined content. The final content of a page is determined only when the visitor requests a page from the web server. Because the final content of the page varies from request to request based on the visitor's actions, this kind of page is called a dynamic page.

Web applications are built to address a variety of challenges and problems. This section describes common uses for web applications and gives a simple example.

- **Web application example**

Janet is a professional web designer and longtime Dreamweaver user responsible for maintaining the intranet and Internet sites of a medium-sized company of 1000 employees. One day, Chris from Human Resources comes to her with a problem. HR administers an employee fitness program that gives employees points for every mile walked, biked, or run. Each employee must report his or her monthly mile totals in an e-mail to Chris. At the end of the month, Chris gathers all the e-mail messages and awards employees small cash prizes according to their point totals.

Chris's problem is that the fitness program has grown too successful. So many employees now participate that Chris is inundated with e-mails at the end of each month. Chris asks Janet if a web-based solution exists.

Janet proposes an intranet-based web application that performs the following tasks:

Lets employees enter their mileage on a web page using a simple HTML form

Stores the employees' mileage in a database

Calculates fitness points based on the mileage data

Lets employees track their monthly progress

Gives Chris one-click access to point totals at the end of each month

● **How a web application works**

A web application is a collection of static and dynamic web pages. A static web page is one that does not change when a site visitor requests it: The web server sends the page to the requesting web browser without modifying it. In contrast, a dynamic web page is modified by the server before it is sent to the requesting browser. The changing nature of the page is why it's called dynamic.

For example, you could design a page to display fitness results, while leaving certain information (such as employee name and results) to be determined when the page is requested by a particular employee.

The next sections describe how web applications work in greater detail.

● **Process static web pages**

A static website comprises a set of related HTML pages and files hosted on a computer running a web server.

A web server is software that serves web pages in response to requests from web browsers. A page request is generated when a visitor clicks a link on a web page, selects a bookmark in a browser, or enters a URL in a browser's address text box.

The final content of a static web page is determined by the page designer and doesn't change when the page is requested. Here's an example:

```
<html>
    <head>
        <title>Trio Motors Information Page</title>
    </head>
    <body>
        <h1>About Trio Motors</h1>
        <p>Trio Motors is a leading automobile manufacturer.</p>
    </body>
</html>
```

Every line of the page's HTML code is written by the designer before the page is placed on the server. Because the HTML doesn't change once it's on the server, this kind of page is called a static page.

Note:

Strictly speaking, a "static" page may not be static at all. For example, a rollover image or Flash content (a SWF file) can make a static page come alive. However, this documentation refers to a page as static if it is sent to the browser without modifications.

When the web server receives a request for a static page, the server reads the request, finds the page, and sends it to the requesting browser, as the following example shows(ref. Fig.1-10-1):

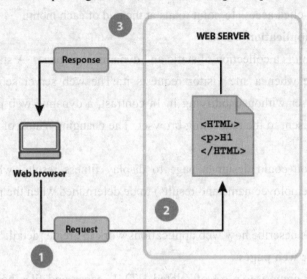

Fig.1-10-1 Process static web page

A. Web browser requests static page.
B. Web server finds page.
C. Web server sends page to requesting browser.

In the case of web applications, certain lines of code are undetermined when the visitor requests the page. These lines must be determined by some mechanism before the page can be sent to the browser. The mechanism is discussed in the following section.

● **Process dynamic pages**

When a web server receives a request for a static web page, the server sends the page directly to the requesting browser. When the web server receives a request for a dynamic page, however, it reacts differently: It passes the page to a special piece of software responsible for finishing the page. This special software is called an application server.

The application server reads the code on the page, finishes the page according to the instructions in the code, and then removes the code from the page. The result is a static page that the application server passes back to the web server, which then sends the page to the requesting browser. All the browser gets when the page arrives is pure HTML. Here's a view of the process(ref. Fig.1-10-2):

A. Web browser requests dynamic page.
B. Web server finds page and passes it to application server.
C. Application server scans page for instructions and finishes page.
D. Application server passes finished page back to web server.
E. Web server sends finished page to requesting browser.

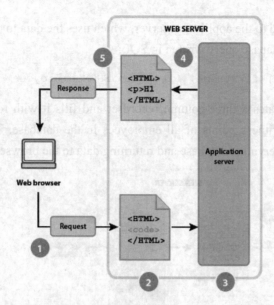

Fig.1-10-2   Process dynamic pages

- **Access a database**

An application server lets you work with server-side resources such as databases. For example, a dynamic page may instruct the application server to extract data from a database and insert it into the page's HTML.

Using a database to store content allows you to separate your website's design from the content you want to display to site users. Instead of writing individual HTML files for every page, you only need to write a page—or template—for the different kinds of information you want to present. You can then upload content into a database and then have the website retrieve that content in response to a user request. You can also update information in a single source, and then populate that change throughout the website without having to manually edit each page. You can use Adobe Dreamweaver to design web forms to insert, update, or delete data from the database.

The instruction to extract data from a database is called a database query. A query consists of search criteria expressed in a database language called SQL (Structured Query Language). The SQL query is written into the page's server-side scripts or tags.

An application server cannot communicate directly with a database because the database's proprietary format renders the data undecipherable in much the same way that a Microsoft Word document opened in Notepad or BBEdit may be undecipherable. The application server can communicate with the database only through the intermediary of a database driver: software that acts like an interpreter between the application server and the database.

After the driver establishes communication, the query is executed against the database and a recordset is created. A recordset is a set of data extracted from one or more tables in a database.

The recordset is returned to the application server, which uses the data to complete the page.

Here's a simple database query written in SQL:

```
SELECT lastname, firstname, fitpoints FROM employees
```

This statement creates a three-column recordset and fills it with rows containing the last name, first name, and fitness points of all employees in the database. The following example shows the process of querying a database and returning data to the browser(ref. Fig.1-10-3):

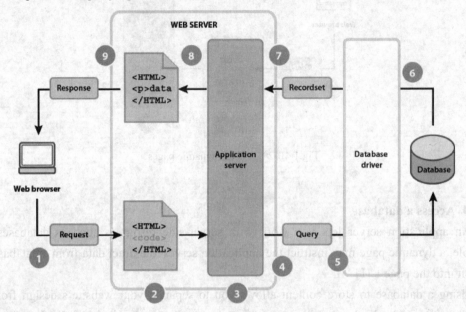

Fig.1-10-3　Access a database

  A. Web browser requests dynamic page.

  B. Web server finds page and passes it to application server.

  C. Application server scans page for instructions.

  D. Application server sends query to database driver.

  E. Driver executes the query against the database.

  F. Recordset is returned to driver.

  G. Driver passes recordset to application server.

  H. Application server inserts data in page, and then passes the page to the web server.

  I. Web server sends finished page to requesting browser.

You can use almost any database with your web application, as long as the appropriate database driver for it is installed on the server.

If you plan to build small low-cost applications, you can use a file-based database, such as one created in Microsoft Access. If you plan to build robust, business-critical applications, you can use a server-based database, such as one created in Microsoft SQL Server, Oracle 9i, or

MySQL.

If your database is located on a system other than your web server, make sure you have a fast connection between the two systems so that your web application can operate quickly and efficiently.

- **Web application terminology**

This section defines frequently used terms relating to web applications.

- **An application server**

Software that helps a web server process web pages containing server-side scripts or tags. When such a page is requested from the server, the web server hands the page off to the application server for processing before sending the page to the browser.

Common application servers include ColdFusion and PHP.

- **A database**

A collection of data stored in tables. Each row of a table constitutes one record and each column constitutes a field in the record, as shown in the following example(ref. Fig.1-10-4):

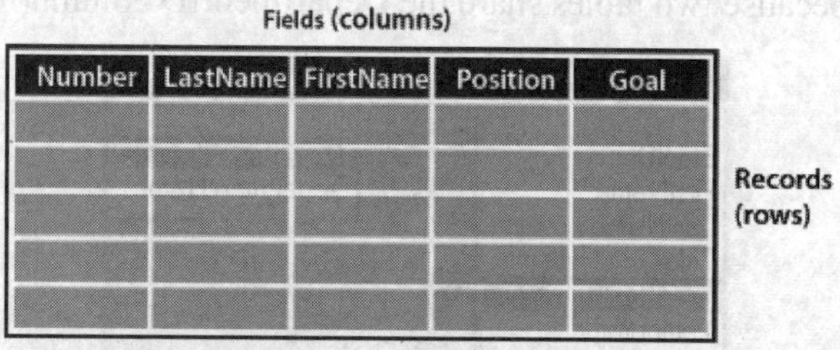

Fig.1-10-4 A database

- **A database driver**

Software that acts as an interpreter between a web application and a database. Data in a database is stored in a proprietary format. A database driver lets the web application read and manipulate data that would otherwise be undecipherable.

- **A database management system**

(DBMS, or database system) Software used to create and manipulate databases. Common database systems include Microsoft Access, Oracle 9i, and MySQL.

- **A database query**

The operation that extracts a recordset from a database. A query consists of search criteria expressed in a database language called SQL. For example, the query can specify that only certain columns or only certain records be included in the recordset.

- **A dynamic page**

A web page customized by an application server before the page is sent to a browser.

- **A recordset**

A set of data extracted from one or more tables in a database, as in the following example(ref. Fig.1-10-5):

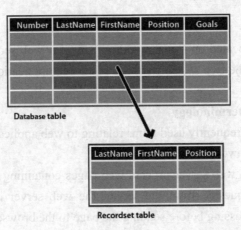

Fig.1-10-5   A recordset

- **A relational database**

A database containing more than one table, with the tables sharing data. The following database is relational because two tables share the DepartmentID column(ref. Fig.1-10-6).

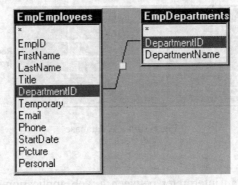

Fig.1-10-6   Relational database

- **A server technology**

The technology that an application server uses to modify dynamic pages at runtime.
The Dreamweaver development environment supports the following server technologies:
Adobe ColdFusion
Microsoft Active Server Pages (ASP)
PHP: Hypertext Preprocessor (PHP)
You can also use the Dreamweaver coding environment to develop pages for any other server technology not listed.

- **A static page**

A web page that is not modified by an application server before the page is sent to a browser.

- **A Web application**

A website that contains pages with partly or entirely undetermined content. The final content of these pages is determined only when a visitor requests a page from the web server. Because the final content of the page varies from request to request based on the visitor's actions, this kind of page is called a dynamic page.

- **A Web server**

Software that sends out web pages in response to requests from web browsers. A page request is generated when a visitor clicks a link on a web page in the browser, selects a bookmark in the browser, or enters a URL in the browser's address text box.

Popular web servers include Microsoft Internet Information Server (IIS) and Apache HTTP Server.

术语参考译文

万维网

Web 或万维网，本质是一个支持特殊格式文档的因特网服务器系统。该文件用称为 HTML（超文本标记语言）的标记语言格式化，HTML 支持指向其他文档，以及图形、音频和视频文件的链接。

这意味着你可以从一个文件跳转到另一个简单的点击热点。并非所有的互联网服务器都是万维网的一部分。

因特网地址

因特网地址唯一标识因特网上的节点。因特网地址也可以指一个网站的名称或 IP（URL）。

超链接

电子文档中的一个元素，它链接到同一文档中的另一个地方或一个完全不同的文档。

超媒体

超媒体是超文本的扩展，除了文本元素，还支持链接图形、声音和视频元素。

网页

万维网上一个文档。每个网页都由一个唯一的 URL（统一资源定位标识）标识。

Java 小程序

applet 是一个用 Java 写的小型互联网程序、一种 Web 编程语言，它可以在任何计算机上下载。

JavaScript

一个由 Netscape 公司开发的脚本语言，供 Web 作者设计互动网站时使用。

CSS 级联样式表

级联样式表的缩写，新的功能被添加到 HTML，在页面如何显示方面，给网站开发者和用户更多的控制。使用 CSS，设计师和用户可以创建定义不同元素的样式表（如标题和链接）显示。这些样式表可以应用到任何网页上。

"级联"一词来源于这样一个事实，即多个样式表可以应用于同一网页。CSS 是由 W3C 开发的。

网站

万维网上的一个站点（位置）。每个网站都包含一个主页，这是用户进入网站时看到的第一个文档。

域名

域名用于识别一个或多个 IP 地址。例如，域名 microsoft.com 代表了大约十几个 IP 地址。

DNS 域名系统

域名系统（或服务或服务器），一种将域名转换为 IP 地址的互联网服务。

URL 统一资源定位符

统一资源定位符（URL）的缩写，它是万维网上文档和其他资源的全局地址。例如，https://www.microsoft.com/zh-cn 是一个 URL。URL 是一种统一资源标识符（URI）。

URL 的第一部分称为协议标识符，它指示要使用什么协议，第二部分称为源名称，它指定资源所在地的 IP 地址或域名。协议标识符和资源名称由冒号和两个正斜杠分隔。

首页

一个网站的主页。通常，主页用于存储在站点中的其他文档的索引或目录。

框架

一个由大多数现代 Web 浏览器支持的功能，使 Web 作者能够将浏览器显示区域分成两个或多个部分（框架）。每个框架的内容取自不同的网页。

标签

在文档中插入的命令，该命令指定文档或文档的一部分应该如何格式化。

网页制作

使用户能够以桌面排版格式开发网站的一类软件。

容器

在 HTML 中，容器是由开始和结束标签所包围的区域。例如<html>包围整个文档，而其他标签可能会包围一个单词、段落或其他元素。

鼠标悬停

一个 JavaScript 元素，当指针经过网页中一个项目时触发该项目的变化（通常是图形的变化，如出现图像或超链接）。变化通常表示该项目链接到相关或附加信息。

静态的

一般是指网络或计算机编程的元素，是固定的、不能行动或变化的元素。静态的相反是动态的。

动态 HTML

指每次查看时 Web 内容都会更改。例如，相同的 URL 可能会调用不同的页面，取决于任意数量的参数，例如：

- 读者的地理位置
- 每天的时间
- 读者浏览的前几页
- 读者的简介

产生动态 HTML 的技术很多，包括 CGI 脚本、服务器端包含（SSI）、cookies、Java、JavaScript 和 ActiveX。

Section II Computer Courses

Unit 1 Adobe Dreamweaver CC

1. Understand the anatomy of a website

Learn how most web pages are made using HTML and CSS.

When you look at a web page, you might wonder how it is made. To display a web page, your Internet browser reads and interprets a special set of instructions. These instructions are written primarily in two languages: HTML and CSS.

The World Wide Web Consortium (W3C) defines the web standards, or specifications, for these languages. It is important to understand web standards to ensure that the different web browsers display your pages accurately. Dreamweaver includes features that allow you to build web pages based on the latest web standards. Let's look at how HTML and CSS work together to make a web page.

- **What is HTML?**

Web pages display content and have an underlying structure defined by HTML. The content can include text, images, links, or even audio and video. Most HTML elements are written using opening <element> and closing </element> tags that tell the browser what type of content to display. The tags are written in a specific order to define the structure of the page. The example below shows three common HTML tags.

- <h1> specifies heading text, good for page titles
- <p> specifies paragraph text, good for body text
- <img> displays an image specified by a src (source file)

Note: The HTML in this example displays in upper case to illustrate the concepts more clearly. However, it is best practice to write HTML tags in lower case when coding your own pages(ref. Fig. 2-1-1).

Fig.2-1-1 HTML

- **What is CSS?**

With HTML you've given a basic structure to your web page, but there's not much customization to the style or layout. Eventually you want to add colors, format text and images, and reposition elements on your page. CSS is a language that works with HTML to apply this style and formatting to your web page. Use CSS to find one or more HTML elements, and then provide additional instructions for what to do with that element (such as change its color or center it on the page).

The example below shows how CSS is used to change the appearance of the three HTML elements.

- **h1** The heading text is center-aligned.
- **p** The paragraph text is red, bold, and center-aligned.
- **img** The image is resized to 150px by 100px (ref. Fig.2-1-2).

Fig.2-1-2   CSS

2. Follow the stages of web design

Whether designing your own site, or working as a team, the process of bringing a website to the web can be deconstructed into manageable phases(ref. Fig.2-1-3).

Fig.2-1-3   Step0

- **Plan your content**

Planning good content is arguably the most important phase in web design. This is also the stage that can take the longest amount of time, but it is worth doing it well. A good content plan will increase the presence and usability of your site as well as speed up web development.

Some key points to keep in mind are to know your audience, short and concise is usually better, think about how others will search for your site, and organize content logically(ref. Fig.2-1-4).

Fig.2-1-4   Step1

- **Wireframe your design**

A wireframe is a blueprint, or an abstract visual representation, of a website that makes it easier for you and your team to collaborate during the design and planning stage. Wireframes can be simple diagrams drawn by hand for quickly getting ideas down wherever you are, or they can be a more complex definition of a website's navigation and functionality.

A variety of wireframing tools are available to you including pen and paper, free or commercial wireframe software apps, or even other Creative Cloud apps such as Photoshop. Choose the method and level of detail that work best for you and your team(ref. Fig.2-1-5).

Fig.2-1-5   Step2

- **Create web pages**

Now it's time to put that great design and planning into action. To bring a conceptual plan to life, it is important to understand the languages of web design. The main focus for this series will be on HTML and CSS with a sneak peek into JavaScript.

Adobe Dreamweaver CC

HTML, or Hypertext Markup Language, is used to create the overall structure and content of the pages in your website. CSS, aka Cascading Style Sheets, is used to style visual properties of website's content and customize the layout. JavaScript is a popular language you can use to add functionality and interactivity to web pages(ref. Fig.2-1-6).

Fig.2-1-6 Step3

- **Publish your website**

Once you are ready to publish your website for the world to see, you'll need a few things. First, you'll have to get a domain name, or web address. You will also need dedicated space on a web server and the connection details—FTP is a common way to connect to a web server. If you are creating a website for your company, you can get this information from your IT web admin. If you are publishing your own site, you can get this information from a web host provider.

You can then use the Sites feature to connect to the web server and publish your site straight from Dreamweaver(ref. Fig.2-1-7).

Fig.2-1-7 Step4

3. The Dreamweaver Workspace Basics Workspace overview

The Dreamweaver workspace lets you view documents and object properties. The workspace also places many of the most common operations in toolbars so that you can quickly make changes to your document (ref. Fig.2-1-8).

Fig.2-1-8 Dreamweaver workspace

A. Application bar   B. Document toolbar   C. Document window
D. Workspace switcher   E. Panels   F. Code View   G. Status bar
H. Tag selector   I. Live View   J. Toolbar

4. Understand Cascading Style Sheet

Use this topic to learn the basic concepts of CSS such as CSS rules, selectors, inheritance, and more. Also, learn how to associate CSS with your web pages in Dreamweaver.

● **About Cascading Style Sheets**

Cascading Style Sheets (CSS) is a collection of formatting rules that control the appearance of content in a web page. Using CSS styles to format a page separates content from presentation. The content of your page—the HTML code—resides in the HTML file, and the CSS rules defining the presentation of the code reside in another file (an external style sheet) or in another part of the HTML document (usually the head section). Separating content from presentation makes it much easier to maintain the appearance of your site from a central location because you don't need to update every property on every page whenever you want to make a change. Separating content from presentation also results in simpler and cleaner HTML code, which provides shorter browser loading times, and simplifies navigation for people with accessibility issues (for example, those using screen readers).

CSS gives you great flexibility and control over the exact appearance of your page. With CSS you can control many text properties including specific fonts and font sizes; bold, italics, underlining, and text shadows; text color and background color; link color and link underlining; and much more. By using CSS to control your fonts, you can also ensure a more consistent treatment of your page layout and appearance in multiple browsers.

In addition to text formatting, you can use CSS to control the format and positioning of block-level elements in a web page. A block-level element is a standalone piece of content, usually separated by a new line in the HTML, and visually formatted as a block. For example, **h1** tags, **p** tags, and **div** tags all produce block-level elements on a web page. You can set margins and borders for block-level elements, position them in a specific location, add background color to them, float text around them, and so on. Manipulating block-level elements is in essence the way you lay out pages with CSS.

- **About CSS rules**

A CSS formatting rule consists of two parts—the selector and the declaration (or in most cases, a block of declarations). The selector is a term (such as **p**, **h1**, a class name, or an id) that identifies the formatted element, and the declaration block defines what the style properties are. In the following example, **h1** is the selector, and everything that falls between the braces (**{ }**) is the declaration block:

```
h1 {
font-size: 16 pixels;
font-family: Helvetica;
font-weight:bold;
}
```

An individual declaration consists of two parts, the property (such as **font-family**) and value (such as **Helvetica**). In the previous CSS rule, a particular style has been created for **h1** tags: the text for all **h1** tags linked to this style will be 16 pixels in size, Helvetica font, and bold.

The style (which comes from a rule, or a collection of rules) resides in a place separate from the actual text it's formatting—usually in an external style sheet, or in the head portion of an HTML document. Thus, one rule for **h1** tags can apply to many tags at once (and in the case of external style sheets, the rule can apply to many tags at once on many different pages). In this way, CSS provides extremely easy update capability. When you update a CSS rule in one place, the formatting of all the elements that use the defined style are automatically updated to the new style(ref. Fig.2-1-9).

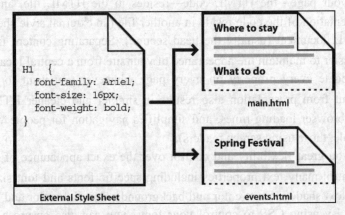

Fig.2-1-9　CSS rules

You can define the following types of styles in Dreamweaver:
- *Class styles* let you apply style properties to any element or elements on the page.
- *HTML tag styles* redefine the formatting for a particular tag, such as `h1`. When you create or change a CSS style for the `h1` tag, all text formatted with the `h1` tag is immediately updated.
- *Advanced styles* redefine the formatting for a particular combination of elements, or for other selector forms as allowed by CSS (for example, the selector `td h2` applies whenever an `h2` header appears inside a table cell.) Advanced styles can also redefine the formatting for tags that contain a specific `id` attribute (for example, the styles defined by `#myStyle` apply to all tags that contain the attribute-value pair `id="myStyle"`).

CSS rules can reside in the following locations:

- **External CSS style sheets**

Collections of CSS rules stored in a separate, external CSS (.css) file (not an HTML file). This file is linked to one or more pages in a website using a link or an @import rule in the head section of a document.

- **Internal (or embedded) CSS style sheets**

Collections of CSS rules included in a `style` tag in the head portion of an HTML document.

- **Inline styles**

Defined within specific instances of tags throughout an HTML document. (Using Inline styles is not recommended.)

Dreamweaver recognizes styles defined in existing documents as long as they conform to CSS style guidelines. Dreamweaver also renders most applied styles directly in Design view. (Previewing the document in a browser window, however, gives you the most accurate "live" rendering of the page.) Some CSS styles are rendered differently in Microsoft Internet Explorer, Netscape, Opera, Apple Safari, or other browsers, and some are not currently supported by any browser.

Note:

To display the O'Reilly CSS reference guide included with Dreamweaver, select Help > Reference and select O'Reilly CSS Reference from the pop-up menu in the Reference panel.

5. Linking

Learn how to set up navigation between your web pages. Link files and documents, update, change, and test links in Dreamweaver.

Before creating a link, make sure you understand how absolute, document-relative, and site root–relative paths work. You can create several types of links in a document:
- A link to another document or to a file, such as a graphic, movie, PDF, or sound file.
- A named anchor link, which jumps to a specific location in a document.

Adobe Dreamweaver CC

- An e-mail link, which creates a new blank e-mail message with the recipient's address already filled in.
- Null and script links, which you use to attach behaviors to an object or to create a link that executes JavaScript code.

You can use the Property inspector and the Point-To-File icon to create links from an image, an object, or text to another document or file.

Dreamweaver creates the links to other pages in your site using document-relative paths. You can also tell Dreamweaver to create new links using site root—relative paths.

6. About Dreamweaver templates

Learn how to use Dreamweaver templates to design a "fixed" page layout and then create documents based on the template that inherit its page layout.

A template is a special type of document that you use to design a "fixed" page layout; you can then create documents based on the template that inherit its page layout. As you design a template, you specify as "editable" which content users can edit in a document based on that template. Templates enable template authors to control which page elements template users—such as writers, graphic artists, or other web developers—can edit. There are several types of template regions the template author can include in a document.

Note:

Templates enable you to control a large design area and reuse complete layouts. If you want to reuse individual design elements, such as a site's copyright information or a logo, create library items.

Using templates enables you to update multiple pages at once. A document that is created from a template remains connected to that template (unless you detach the document later). You can modify a template and immediately update the design in all documents based on it.

Unit 2 C Programming

1. C—Program Structure

Before we study the basic building blocks of the C programming language, let us look at a bare minimum C program structure so that we can take it as a reference in the upcoming chapters.

Hello World Example

A C program basically consists of the following parts.
- Preprocessor Commands
- Functions
- Variables
- Statements & Expressions
- Comments

Let us look at a simple code that would print the words "Hello, World!".

```
#include <stdio.h>
int main()
{   /* my first program in C */
    printf("Hello, World! \n");
       return 0;
}
```

Let us take a look at the various parts of the above program.
- The first line of the program #include <stdio.h> is a preprocessor command, which tells a C compiler to include stdio.h file before going to actual compilation.
- The next line int main() is the main function where the program execution begins.
- The next line /*...*/ will be ignored by the compiler and it has been put to add additional comments in the program. So such lines are called comments in the program.
- The next line printf(...) is another function available in C which causes the message "Hello, World!" to be displayed on the screen.
- The next line return 0; terminates the main() function and returns the value 0.

2. C—Loops

You may encounter situations, when a block of code needs to be executed several number of times. In general, statements are executed sequentially: The first statement in a function is executed first, followed by the second, and so on.

Programming languages provide various control structures that allow for more complicated execution paths.

A loop statement allows us to execute a statement or group of statements multiple times. Given below is the general form of a loop statement in most of the programming languages (ref. Fig.2-2-1).

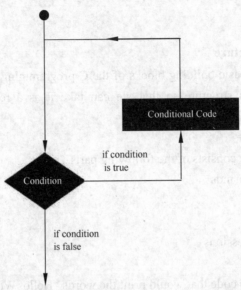

Fig. 2-2-1   Form of a loop statement

C programming language provides the following types of loops to handle looping requirements(ref. Table 2-2-1).

Table 2-2-1   Loop type & description

S.N.	Loop type & description
1	while loop Repeats a statement or group of statements while a given condition is true. It tests the condition before executing the loop body
2	for loop Executes a sequence of statements multiple times and abbreviates the code that manages the loop variable
3	do…while loop It is more like a while statement, except that it tests the condition at the end of the loop body
4	nested loops You can use one or more loops inside any other while, for, or do…while loop

● **Loop control statements**

Loop control statements change execution from its normal sequence. When execution leaves a scope, all automatic objects that were created in that scope are destroyed.

C supports the following control statements(ref. Table 2-2-2).

Table 2-2-2 Control statement & description

S.N.	Control statement & description
1	break statement Terminates the loop or switch statement and transfers execution to the statement immediately following the loop or switch.
2	continue statement Causes the loop to skip the remainder of its body and immediately retest its condition prior to reiterating.
3	goto statement Transfers control to the labeled statement.

- **The infinite loop**

A loop becomes an infinite loop if a condition never becomes false. The for loop is traditionally used for this purpose. Since none of the three expressions that form the "for" loop are required, you can make an endless loop by leaving the conditional expression empty.

```
#include <stdio.h>
 int main ()
{  for( ; ; )
{  printf("This loop will run forever.\n");
}
   return 0;
}
```

When the conditional expression is absent, it is assumed to be true. You may have an initialization and increment expression, but C programmers more commonly use the for(;;) construct to signify an infinite loop.

3. C—Functions

A function is a group of statements that together perform a task. Every C program has at least one function, which is main(), and all the most trivial programs can define additional functions.

You can divide up your code into separate functions. How you divide up your code among different functions is up to you, but logically the division is such that each function performs a specific task.

A function declaration tells the compiler about a function's name, return type, and parameters. A function definition provides the actual body of the function.

The C standard library provides numerous built-in functions that your program can call. For example, strcat() to concatenate two strings, memcpy() to copy one memory location to another location, and many more functions.

A function can also be referred as a method or a sub-routine or a procedure, etc.

- **Defining a function**

The general form of a function definition in C programming language is as follows—

```
return_type function_name( parameter list ) {
   body of the function
}
```

A function definition in C programming consists of a function header and afunction body. Here are all the parts of a function—

- Return Type—A function may return a value. The return_type is the data type of the value the function returns. Some functions perform the desired operations without returning a value. In this case, the return_type is the keyword void.
- Function Name—This is the actual name of the function. The function name and the parameter list together constitute the function signature.
- Parameters—A parameter is like a placeholder. When a function is invoked, you pass a value to the parameter. This value is referred to as actual parameter or argument. The parameter list refers to the type, order, and number of the parameters of a function. Parameters are optional; that is, a function may contain no parameters.
- Function Body—The function body contains a collection of statements that define what the function does.

● **Example**

Given below is the source code for a function called max(). This function takes two parameters num1 and num2 and returns the maximum value between the two—

```
/* function returning the max between two numbers */
int max(int num1, int num2)
{ /* local variable declaration */
   int result;
   if (num1 > num2)
      result = num1;
   else
      result = num2;
   return result;
}
```

● **Function declarations**

A function declaration tells the compiler about a function name and how to call the function. The actual body of the function can be defined separately.

A function declaration has the following parts—

```
return_type function_name( parameter list );
```

For the above defined function max(), the function declaration is as follows—

```
int max(int num1, int num2);
```

Parameter names are not important in function declaration only their type is required, so the following is also a valid declaration—

```
int max(int, int);
```

Function declaration is required when you define a function in one source file and you call that function in another file. In such case, you should declare the function at the top of the file calling the function.

- **Calling a function**

While creating a C function, you give a definition of what the function has to do. To use a function, you will have to call that function to perform the defined task.

When a program calls a function, the program control is transferred to the called function. A called function performs a defined task and when its return statement is executed or when its function-ending closing brace is reached, it returns the program control back to the main program.

To call a function, you simply need to pass the required parameters along with the function name, and if the function returns a value, then you can store the returned value. For example—

```
#include <stdio.h>
 /* function declaration */
int max(int num1, int num2);
 int main ()
{ /* local variable definition */
   int a = 100;
   int b = 200;
   int ret;
   /* calling a function to get max value */
  ret = max(a, b);
   printf( "Max value is : %d\n", ret );
   return 0;
}

/* function returning the max between two numbers */
int max(int num1, int num2)
{ /* local variable declaration */
   int result;
   if (num1 > num2)
     result = num1;
   else
     result = num2;
   return result;
}
```

We have kept max() along with main() and compiled the source code. While running the final executable, it would produce the following result—

```
Max value is : 200
```

- **Function arguments**

If a function is to use arguments, it must declare variables that accept the values of the arguments. These variables are called the formal parameters of the function.

Formal parameters behave like other local variables inside the function and are created upon entry into the function and destroyed upon exit.

While calling a function, there are two ways in which arguments can be passed to a function(ref. Table 2-2-3).

Table 2-2-3 Call type & description

S.N.	Call type & description
1	Call by value This method copies the actual value of an argument into the formal parameter of the function. In this case, changes made to the parameter inside the function have no effect on the argument.
2	Call by reference This method copies the address of an argument into the formal parameter. Inside the function, the address is used to access the actual argument used in the call. This means that changes made to the parameter affect the argument.

By default, C uses call by value to pass arguments. In general, it means the code within a function cannot alter the arguments used to call the function.

4. C—Arrays

Arrays a kind of data structure that can store a fixed-size sequential collection of elements of the same type. An array is used to store a collection of data, but it is often more useful to think of an array as a collection of variables of the same type.

Instead of declaring individual variables, such as number0, number1, ..., and number99, you declare one array variable such as numbers and use numbers[0], numbers[1], and ..., numbers[99] to represent individual variables. A specific element in an array is accessed by an index.

All arrays consist of contiguous memory locations. The lowest address corresponds to the first element and the highest address to the last element(ref. Fig.2-2-2).

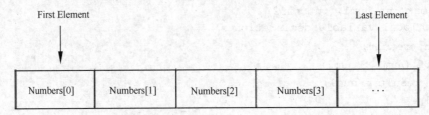

Fig.2-2-2　C array

- **Declaring arrays**

To declare an array in C, a programmer specifies the type of the elements and the number of elements required by an array as follows—

```
type arrayName [ arraySize ];
```

This is called a single-dimensional array. The arraySize must be an integer constant greater than zero and type can be any valid C data type. For example, to declare a 10-element array called balance of type double, use this statement—

```
double balance[10];
```

Here balance is a variable array which is sufficient to hold up to 10 double numbers.

- **Initializing arrays**

You can initialize an array in C either one by one or using a single statement as follows —

```
double balance[5] = {1000.0, 2.0, 3.4, 7.0, 50.0};
```

The number of values between braces { } cannot be larger than the number of elements that we declare for the array between square brackets [ ].

If you omit the size of the array, an array just big enough to hold the initialization is created. Therefore, if you write—

```
double balance[] = {1000.0, 2.0, 3.4, 7.0, 50.0};
```

You will create exactly the same array as you did in the previous example. Following is an example to assign a single element of the array—

```
balance[4] = 50.0;
```

The above statement assigns the 5th element in the array with a value of 50.0. All arrays have 0 as the index of their first element which is also called the base index and the last index of an array will be total size of the array minus 1. Shown below is the pictorial representation of the array we discussed above(ref. Fig.2-2-3).

	0	1	2	3	4
balance	1000.0	2.0	3.4	7.0	50.0

Fig.2-2-3　Initialize an array

- **Accessing array elements**

An element is accessed by indexing the array name. This is done by placing the index of the element within square brackets after the name of the array. For example—

```
double salary = balance[9];
```

The above statement will take the 10th element from the array and assign the value to salary variable. The following example Shows how to use all the three above mentioned concepts viz.

declaration, assignment, and accessing arrays—

```
#include <stdio.h>
 int main ()
{ int n[ 10 ]; /* n is an array of 10 integers */
   int i,j;
    /* initialize elements of array n to 0 */
   for ( i = 0; i < 10; i++ )
{ n[ i ] = i + 100; /* set element at location i to i + 100 */
   }

   /* output each array element's value */
   for (j = 0; j < 10; j++ )
{ printf("Element[%d] = %d\n", j, n[j] );
   }
    return 0;
}
```

When the above code is compiled and executed, it produces the following result—

```
Element[0] = 100
Element[1] = 101
Element[2] = 102
Element[3] = 103
Element[4] = 104
Element[5] = 105
Element[6] = 106
Element[7] = 107
Element[8] = 108
Element[9] = 109
```

- **Arrays in detail**

Arrays are important to C and should need a lot more attention. The following important concepts related to array should be clear to a C programmer(ref. Table 2-2-4).

Table 2-2-4 Array concept & description

S.N.	Array concept & description
1	Multi-dimensional arrays C supports multidimensional arrays. The simplest form of the multidimensional array is the two-dimensional array.
2	Passing arrays to functions You can pass to the function a pointer to an array by specifying the array's name without an index.
3	Return array from a function C allows a function to return an array.
4	Pointer to an array You can generate a pointer to the first element of an array by simply specifying the array name, without any index.

5. C—Pointers

Pointers in C are easy and fun to learn. Some C programming tasks are performed more easily with pointers, and other tasks, such as dynamic memory allocation, cannot be performed without using pointers. So it becomes necessary to learn pointers to become a perfect C programmer. Let's start learning them in simple and easy steps.

As you know, every variable is a memory location and every memory location has its address defined which can be accessed using ampersand (&) operator, which denotes an address in memory. Consider the following example, which prints the address of the variables defined—

```
#include <stdio.h>
int main ()
{   int var1;
    char var2[10];
    printf("Address of var1 variable: %x\n", &var1 );
    printf("Address of var2 variable: %x\n", &var2 );
    return 0;
}
```

When the above code is compiled and executed, it produces the following result—

```
Address of var1 variable: bff5a400
Address of var2 variable: bff5a3f6
```

● **What are pointers?**

A pointer is a variable whose value is the address of another variable, i.e., direct address of the memory location. Like any variable or constant, you must declare a pointer before using it to store any variable address. The general form of a pointer variable declaration is—

```
type *var-name;
```

Here, type is the pointer's base type; it must be a valid C data type and var-name is the name of the pointer variable. The asterisk * used to declare a pointer is the same asterisk used for multiplication. However, in this statement the asterisk is being used to designate a variable as a pointer. Take a look at some of the valid pointer declarations—

```
int    *ip;     /* pointer to an integer */
double *dp;     /* pointer to a double */
float  *fp;     /* pointer to a float */
char   *ch;     /* pointer to a character */
```

The actual data type of the value of all pointers, whether integer, float, character, or otherwise, is the same, a long hexadecimal number that represents a memory address. The only difference between pointers of different data types is the data type of the variable or constant that the pointer points to.

- **How to use pointers?**

There are a few important operations, which we will do with the help of pointers very frequently. (a) We define a pointer variable, (b) assign the address of a variable to a pointer and (c) finally access the value at the address available in the pointer variable. This is done by using unary operator * that returns the value of the variable located at the address specified by its operand. The following example makes use of these operations—

```
#include <stdio.h>
int main ()
{  int var = 20;     /* actual variable declaration */
   int *ip;          /* pointer variable declaration */
   ip = &var;        /* store address of var in pointer variable*/
   printf("Address of var variable: %x\n", &var );

   /* address stored in pointer variable */
   printf("Address stored in ip variable: %x\n", ip );
   /* access the value using the pointer */
   printf("Value of *ip variable: %d\n", *ip );
   return 0;
}
```

When the above code is compiled and executed, it produces the following result.

```
Address of var variable: bffd8b3c
Address stored in ip variable: bffd8b3c
Value of *ip variable: 20
```

- **NULL pointers**

It is always a good practice to assign a NULL value to a pointer variable in case you do not have an exact address to be assigned. This is done at the time of variable declaration. A pointer that is assigned NULL is called a null pointer.

The NULL pointer is a constant with a value of zero defined in several standard libraries. Consider the following program.

```
#include <stdio.h>
int main ()
{  int *ptr = NULL;
   printf("The value of ptr is : %x\n", ptr );
    return 0;
}
```

When the above code is compiled and executed, it produces the following result.

```
The value of ptr is 0
```

In most of the operating systems, programs are not permitted to access memory at address 0 because that memory is reserved by the operating system. However, the memory address 0 has special significance; it signals that the pointer is not intended to point to an accessible memory location. But by convention, if a pointer contains the null (zero) value, it is assumed to point to nothing.

To check for a null pointer, you can use an "if" statement as follows.

```
if(ptr)     /* succeeds if p is not null */
if(!ptr)    /* succeeds if p is null */
```

- **Pointers in detail**

Pointers have many but easy concepts and they are very important to C programming. The following important pointer concepts should be clear to any C programmer(ref. Table 2-2-5).

Table 2-2-5 Pointer concept & description

S.N.	Pointer concept & description
1	Pointer arithmetic There are four arithmetic operators that can be used in pointers: ++, --, +, -
2	Array of pointers You can define arrays to hold a number of pointers.
3	Pointer to pointer C allows you to have pointer on a pointer and so on.
4	Passing pointers to functions in C Passing an argument by reference or by address enable the passed argument to be changed in the calling function by the called function.
5	Return pointer from functions in C C allows a function to return a pointer to the local variable, static variable, and dynamically allocated memory as well.

6. C—Strings

Strings are actually one-dimensional array of characters terminated by a nullcharacter '\0'. Thus a null-terminated string contains the characters that comprise the string followed by a null.

The following declaration and initialization create a string consisting of the word "Hello". To hold the null character at the end of the array, the size of the character array containing the string is one more than the number of characters in the word "Hello".

```
char greeting[6] = {'H', 'e', 'l', 'l', 'o', '\0'};
```

If you follow the rule of array initialization then you can write the above statement as follows.

```
char greeting[] = "Hello";
```

Following is the memory presentation of the above defined string in C/C++ (ref. Fig.2-2-4).

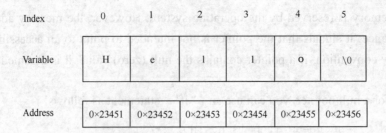

Fig.2-2-4　String memory presentation

Actually, you do not place the null character at the end of a string constant. The C compiler automatically places the '\0' at the end of the string when it initializes the array. Let us try to print the above mentioned string.

```
#include <stdio.h>
int main ()
{   char greeting[6] = {'H', 'e', 'l', 'l', 'o', '\0'};
    printf("Greeting message: %s\n", greeting );
    return 0;
}
```

When the above code is compiled and executed, it produces the following result—

```
Greeting message: Hello
```

C supports a wide range of functions that manipulate null-terminated strings (ref. Table 2-2-6).

Table 2-2-6　Function & purpose

S.N.	Function & purpose
1	strcpy(s1, s2); Copies string s2 into string s1.
2	strcat(s1, s2); Concatenates string s2 onto the end of string s1.
3	strlen(s1); Returns the length of string s1.
4	strcmp(s1, s2); Returns 0 if s1 and s2 are the same; less than 0 if s1<s2; greater than 0 if s1>s2.
5	strchr(s1, ch); Returns a pointer to the first occurrence of character ch in string s1.
6	strstr(s1, s2); Returns a pointer to the first occurrence of string s2 in string s1.

The following example uses some of the above-mentioned functions.

```
#include <stdio.h>
#include <string.h>
int main ()
{   char str1[12] = "Hello";
    char str2[12] = "World";
    char str3[12];
    int  len ;
```

```
    /* copy str1 into str3 */
    strcpy(str3, str1);
    printf("strcpy( str3, str1) :  %s\n", str3 );

    /* concatenates str1 and str2 */
    strcat( str1, str2);
    printf("strcat( str1, str2):   %s\n", str1 );

    /* total length of str1 after concatenation */
    len = strlen(str1);
    printf("strlen(str1) :  %d\n", len );
    return 0;
}
```

When the above code is compiled and executed, it produces the following result—

```
strcpy( str3, str1) :   Hello
strcat( str1, str2):    HelloWorld
strlen(str1) :  10
```

7. C—Structures

Arrays allow to define type of variables that can hold several data items of the same kind. Similarly structure is another user defined data type available in C that allows to combine data items of different kinds.

Structures are used to represent a record. Suppose you want to keep track of your books in a library. You might want to track the following attributes about each book.

(1) Title
(2) Author
(3) Subject
(4) Book ID

● **Defining a structure**

To define a structure, you must use the struct statement. The struct statement defines a new data type, with more than one member. The format of the struct statement is as follows.

```
struct [structure tag] {
  member definition;
  member definition;
  ...
  member definition;
} [one or more structure variables];
```

The structure tag is optional and each member definition is a normal variable definition, such as int i; or float f; or any other valid variable definition. At the end of the structure's definition, before the final semicolon, you can specify one or more structure variables but it is optional. Here is the way you would declare the Book structure.

```
struct Books {
   char  title[50];
   char  author[50];
   char  subject[100];
   int   book_id;
} book;
```

- **Accessing structure members**

To access any member of a structure, we use the member access operator (.). The member access operator is coded as a period between the structure variable name and the structure member that we wish to access. You would use the keyword struct to define variables of structure type. The following example shows how to use a structure in a program.

```
#include <stdio.h>
#include <string.h>
struct Books {
char  title[50];
   char  author[50];
   char  subject[100];
   int   book_id;
};

int main( )
 {struct Books Book1;        /* Declare Book1 of type Book */
  struct Books Book2;        /* Declare Book2 of type Book */

   /* book 1 specification */
   strcpy( Book1.title, "C Programming");
   strcpy( Book1.author, "Nuha Ali");
   strcpy( Book1.subject, "C Programming Tutorial");
   Book1.book_id = 6495407;

   /* book 2 specification */
   strcpy( Book2.title, "Telecom Billing");
   strcpy( Book2.author, "Zara Ali");
   strcpy( Book2.subject, "Telecom Billing Tutorial");
   Book2.book_id = 6495700;

   /* print Book1 info */
   printf( "Book 1 title : %s\n", Book1.title);
   printf( "Book 1 author : %s\n", Book1.author);
   printf( "Book 1 subject : %s\n", Book1.subject);
```

```
    printf( "Book 1 book_id : %d\n", Book1.book_id);

    /* print Book2 info */
    printf( "Book 2 title : %s\n", Book2.title);
    printf( "Book 2 author : %s\n", Book2.author);
    printf( "Book 2 subject : %s\n", Book2.subject);
    printf( "Book 2 book_id : %d\n", Book2.book_id);
    return 0;
}
```

When the above code is compiled and executed, it produces the following result.

```
Book 1 title : C Programming
Book 1 author : Nuha Ali
Book 1 subject : C Programming Tutorial
Book 1 book_id : 6495407
Book 2 title : Telecom Billing
Book 2 author : Zara Ali
Book 2 subject : Telecom Billing Tutorial
Book 2 book_id : 6495700
```

- **Structures as function arguments**

You can pass a structure as a function argument in the same way as you pass any other variable or pointer.

```
#include <stdio.h>
#include <string.h>
 struct Books {
   char  title[50];
   char  author[50];
   char  subject[100];
   int   book_id;
};

/* function declaration */
void printBook( struct Books book );

int main( )
{  struct Books Book1;       /* Declare Book1 of type Book */
   struct Books Book2;       /* Declare Book2 of type Book */

   /* book 1 specification */
   strcpy( Book1.title, "C Programming");
   strcpy( Book1.author, "Nuha Ali");
   strcpy( Book1.subject, "C Programming Tutorial");
   Book1.book_id = 6495407;
```

```
    /* book 2 specification */
    strcpy( Book2.title, "Telecom Billing");
    strcpy( Book2.author, "Zara Ali");
    strcpy( Book2.subject, "Telecom Billing Tutorial");
    Book2.book_id = 6495700;

    /* print Book1 info */
    printBook( Book1 );

    /* Print Book2 info */
    printBook( Book2 );
    return 0;
}

void printBook( struct Books book )
{   printf( "Book title : %s\n", book.title);
    printf( "Book author : %s\n", book.author);
    printf( "Book subject : %s\n", book.subject);
    printf( "Book book_id : %d\n", book.book_id);
}
```

When the above code is compiled and executed, it produces the following result.

```
Book title : C Programming
Book author : Nuha Ali
Book subject : C Programming Tutorial
Book book_id : 6495407
Book title : Telecom Billing
Book author : Zara Ali
Book subject : Telecom Billing Tutorial
Book book_id : 6495700
```

- **Pointers to structures**

You can define pointers to structures in the same way as you define pointer to any other variable.

```
struct Books *struct_pointer;
```

Now, you can store the address of a structure variable in the above defined pointer variable. To find the address of a structure variable, place the '&'; operator before the structure's name as follows.

```
struct_pointer = &Book1;
```

To access the members of a structure using a pointer to that structure, you must use the operator as follows.

```
struct_pointer->title;
```

Let us re-write the above example using structure pointer.

```
#include <stdio.h>
#include <string.h>
 struct Books {
   char  title[50];
   char  author[50];
   char  subject[100];
   int   book_id;
};

/* function declaration */
void printBook( struct Books *book );
int main( )
{  struct Books Book1;        /* Declare Book1 of type Book */
   struct Books Book2;        /* Declare Book2 of type Book */

   /* book 1 specification */
   strcpy( Book1.title, "C Programming");
   strcpy( Book1.author, "Nuha Ali");
   strcpy( Book1.subject, "C Programming Tutorial");
   Book1.book_id = 6495407;

   /* book 2 specification */
   strcpy( Book2.title, "Telecom Billing");
   strcpy( Book2.author, "Zara Ali");
   strcpy( Book2.subject, "Telecom Billing Tutorial");
   Book2.book_id = 6495700;

   /* print Book1 info by passing address of Book1 */
   printBook( &Book1 );

   /* print Book2 info by passing address of Book2 */
   printBook( &Book2 );
   return 0;
}

void printBook( struct Books *book )
{  printf( "Book title : %s\n", book->title);
   printf( "Book author : %s\n", book->author);
   printf( "Book subject : %s\n", book->subject);
   printf( "Book book_id : %d\n", book->book_id);
```

}

When the above code is compiled and executed, it produces the following result—

```
Book title : C Programming
Book author : Nuha Ali
Book subject : C Programming Tutorial
Book book_id : 6495407
Book title : Telecom Billing
Book author : Zara Ali
Book subject : Telecom Billing Tutorial
Book book_id : 6495700
```

8. C—Input & Output

When we say Input, it means to feed some data into a program. An input can be given in the form of a file or from the command line. C programming provides a set of built-in functions to read the given input and feed it to the program as per requirement.

When we say Output, it means to display some data on screen, printer, or in any file. C programming provides a set of built-in functions to output the data on the computer screen as well as to save it in text or binary files.

- **The standard files**

C programming treats all the devices as files. So devices such as the display are addressed in the same way as files and the following three files are automatically opened when a program executes to provide access to the keyboard and screen(ref. Table 2-2-7).

Table 2-2-7 Standard file

Standard file	File pointer	Device
Standard input	stdin	Keyboard
Standard output	stdout	Screen
Standard error	stderr	Your screen

The file pointers are the means to access the file for reading and writing purpose. This section explains how to read values from the screen and how to print the result on the screen.

- **The getchar() and putchar() functions**

The int getchar(void) function reads the next available character from the screen and returns it as an integer. This function reads only single character at a time. You can use this method in the loop in case you want to read more than one character from the screen.

The int putchar(int c) function puts the passed character on the screen and returns the same character. This function puts only single character at a time. You can use this method in the loop in case you want to display more than one character on the screen. Check the following example.

```
#include <stdio.h>
int main( )
{   int c;
```

```
    printf( "Enter a value :");
    c = getchar( );
    printf( "\n You entered: ");
    putchar( c );
    return 0;
}
```

When the above code is compiled and executed, it waits for you to input some text. When you enter a text and press enter, then the program proceeds and reads only a single character and displays it as follows.

```
Enter a value : this is test
You entered: t
```

- **The gets() and puts() functions**

The char *gets(char *s) function reads a line from stdin into the buffer pointed to by s until either a terminating newline or EOF (End of File).

The int puts(const char *s) function writes the string 's' and 'a' trailing newline to stdout.

```
#include <stdio.h>
int main( ) {
    char str[100];
    printf( "Enter a value :");
    gets( str );
    printf( "\n You entered: ");
    puts( str );
    return 0;
}
```

When the above code is compiled and executed, it waits for you to input some text. When you enter a text and press enter, then the program proceeds and reads the complete line till end, and displays it as follows.

```
Enter a value : this is test
You entered: this is test
```

- **The scanf() and printf() functions**

The int scanf(const char *format, ...) function reads the input from the standard input stream stdin and scans that input according to the formatprovided.

The int printf(const char *format, ...) function writes the output to the standard output stream stdout and produces the output according to the format provided.

The format can be a simple constant string, but you can specify %s, %d, %c, %f, etc., to print or read strings, integer, character or float respectively. There are many other formatting options available which can be used based on requirements. Let us now proceed with a simple example to understand the concepts better.

```
#include <stdio.h>
int main( )
{   char str[100];
    int i;
    printf( "Enter a value :");
    scanf("%s %d", str, &i);
    printf( "\n You entered: %s %d ", str, i);
    return 0;
}
```

When the above code is compiled and executed, it waits for you to input some text. When you enter a text and press enter, then program proceeds and reads the input and displays it as follows.

```
Enter a value : seven 7
You entered: seven 7
```

Here, it should be noted that scanf() expects input in the same format as you provided %s and %d, which means you have to provide valid inputs like "string integer". If you provide "string string" or"integer integer", then it will be assumed as wrong input. Secondly, while reading a string, scanf() stops reading as soon as it encounters a space, so "this is test" are three strings for scanf().

Unit 3 Features Available only on Windows 10

- **Sticky Notes with Cortana**

Sticky Notes enable you to capture and save a brilliant thought or jot down an important detail. Now integrated with Cortana, Sticky Notes lets you set reminders that flow across your devices(ref. Fig.2-3-1).

Fig.2-3-1 Sticky Notes

- **Dark mode**

Instantly change your apps from light to dark mode—great for low-light conditions like working at night or on a plane(ref. Fig.2-3-2).

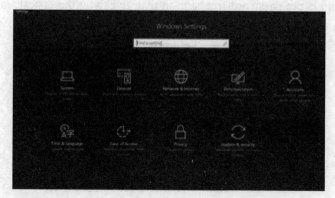

Fig.2-3-2 Dark mode

- **Music controls in lock screen**

If you have music playing when you lock your device, you can now control playback right from your lock screen. Keep listening without missing a beat(ref. Fig.2-3-3).

Fig.2-3-3　Music controls in lock screen

- **The quickest way to set ideas in motion**

Quickly access the Windows Ink Workspace to use Sticky Notes, Sketchpad, and Screen Sketch. Naturally create, capture and edit ideas at the speed of thought within Office (ref. Fig.2-3-4).

Fig.2-3-4　Set ideas in motion

- **Hello, you are the password**

Windows Hello is the password-free sign-in that gives you the fastest, most secure way to unlock your Windows devices(ref. Fig.2-3-5).

Fig.2-3-5　Windows Hello

- **Go beyond browsing**

Microsoft Edge is the browser built for Windows 10. Go beyond browsing—enjoy a web experience that's personal, responsive and all about getting things done online(ref. Fig.2-3-6).

Fig.2-3-6　Microsoft Edge

- **Your truly personal digital assistant**

By learning over time, Cortana becomes more useful as you go. Count on Cortana to help you find things, complete tasks, set reminders and work across your devices more productively(ref. Fig.2-3-7).

Fig.2-3-7　Personal digital assistant

- **The best Windows ever for gaming**

Play great new Xbox games on Windows 10 in native 4K resolution**. Stay connected to your gaming community with the Xbox app. And play where you want with in-home streaming and Xbox Play Anywhere(ref. Fig.2-3-8).

Fig.2-3-8　Xbox games

- **Best apps built for doing**

Get your favorites fast in the Windows Store, your one-stop shop on your PC, tablet, phone and Xbox One. Easily find and acquire popular free and paid apps, desktop software, PC and

Features Available only on Windows 10

Xbox games, movies, TV shows and the latest music(ref. Fig.2-3-9).

Fig.2-3-9　Windows Store

- **Most comprehensive security**

As the most secure Windows ever built, Windows 10 delivers comprehensive protection—including anti-virus, firewall, Windows Defender and anti-phishing technologies—all delivered built-in at no extra cost to you(ref. Fig.2-3-10).

Fig.2-3-10　Secure Windows

- **Emoji keyboard**

The new emoji keyboard features a new font and categories of expressive and playful emoji, with more fidelity and skin tone options; available for 25 languages(ref. Fig.2-3-11).

Fig.2-3-11　Emoji keyboard

- **One place for taking action**

Action Center gives you quick access to the things you need—like notifications from email, apps, Cortana, and Defender—all in a single, easily accessible spot on the right side of your taskbar(ref. Fig.2-3-12).

Fig.2-3-12 Action Center

● **Use your laptop like a tablet**

Tablet Mode delivers a smooth, touch-first experience on your tablet, 2-in-1 or touchscreen laptop. Apps scale smoothly and onscreen features adapt for easy navigation(ref. Fig.2-3-13).

Fig.2-3-13 Tablet Mode

Unit 4 Excel Training

1. Basic tasks in Excel 2016 for Windows

Excel is an incredibly powerful tool for getting meaning out of vast amounts of data. But it also works really well for simple calculations and tracking almost any kind of information. The key for unlocking all that potential is the grid of cells. Cells can contain numbers, text, or formulas. You put data in your cells and group them in rows and columns. That allows you to add up your data, sort and filter it, put it in tables, and build great-looking charts. Let's go through the basic steps to get you started.

- **Create a new workbook**

Excel documents are called workbooks. Each workbook has sheets, typically called spreadsheets. You can add as many sheets as you want to a workbook, or you can create new workbooks to keep your data separate.

(1) Click File, and then click New.

(2) Under New, click the Blank workbook(ref. Fig.2-4-1).

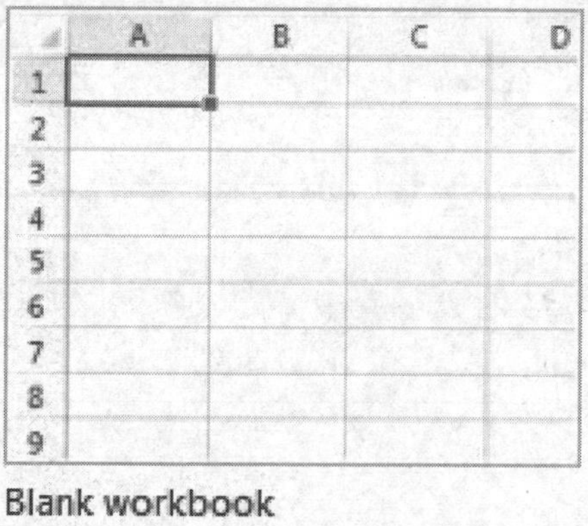

Fig.2-4-1   Blank workbook

- **Enter your data**

(1) Click an empty cell.

For example, cell A1 on a new sheet. Cells are referenced by their location in the row and column on the sheet, so cell A1 is in the first row of column A.

(2) Type text or a number in the cell.

(3) Press Enter or Tab to move to the next cell.

- **Use AutoSum to add your data**

When you've entered numbers in your sheet, you might want to add them up. A fast way to do that is by using AutoSum.

(1) Select the cell to the right or below the numbers you want to add.

(2) Click the Home tab, and then click AutoSum in the Editing group(ref. Fig.2-4-2).

Fig.2-4-2   AutoSum

AutoSum adds up the numbers and shows the result in the cell you selected.

- **Create a simple formula**

Adding numbers is just one of the things you can do, but Excel can do other math as well. Try some simple formulas to add, subtract, multiply, or divide your numbers.

(1) Pick a cell, and then type an equal sign (=).

That tells Excel that this cell will contain a formula.

(2) Type a combination of numbers and calculation operators, like the plus sign (+) for addition, the minus sign (−) for subtraction, the asterisk (*) for multiplication, or the forward slash (/) for division.

For example, enter =2+4, =4−2, =2*4, or =4/2.

(3) Press Enter.

This runs the calculation.

You can also press Ctrl+Enter if you want the cursor to stay on the active cell.

- **Apply a number format**

To distinguish between different types of numbers, add a format, like currency, percentages, or dates.

(1) Select the cells that have numbers you want to format.

(2) Click the Home tab, and then click the arrow in the General box(ref. Fig.2-4-3).

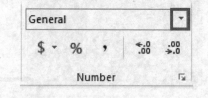

Fig.2-4-3   General box

(3) Pick a number format(ref. Fig.2-4-4).

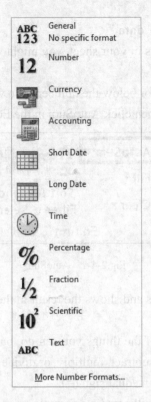

Fig.2-4-4　Number format

If you don't see the number format you're looking for, click More Number Formats.

● **Put your data in a table**

A simple way to access Excel's power is to put your data in a table. That lets you quickly filter or sort your data.

(1) Select your data by clicking the first cell and dragging to the last cell in your data.

To use the keyboard, hold down Shift while you press the arrow keys to select your data.

(2) Click the Quick Analysis button in the bottom-right corner of the selection (ref. Fig.2-4-5).

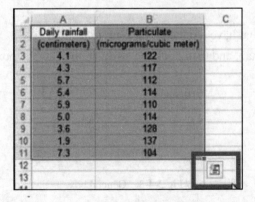

Fig.2-4-5　Quick Analysis button

(3) Click Tables, move your cursor to the Table button to preview your data, and then click the Table button(ref. Fig.2-4-6).

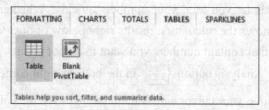

Fig.2-4-6   Table button

(4) Click the arrow ▼ in the table header of a column.

(5) To filter the data, clear the Select All check box, and then select the data you want to show in your table(ref. Fig.2-4-7).

Fig.2-4-7   Filter the data

(6) To sort the data, click Sort A to Z or Sort Z to A(ref. Fig.2-4-8).

Fig.2-4-8   Sort the data

(7) Click OK.

● **Show totals for your numbers**

Quick Analysis tools let you total your numbers quickly. Whether it's a sum, average, or count you want, Excel shows the calculation results right below or next to your numbers.

(1) Select the cells that contain numbers you want to add or count.

(2) Click the Quick Analysis button in the bottom-right corner of the selection.

(3) Click Totals, move your cursor across the buttons to see the calculation results for your data, and then click the button to apply the totals(ref. Fig.2-4-9).

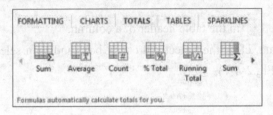

Fig.2-4-9　Quick Analysis tools

● **Add meaning to your data**

Conditional formatting or sparklines can highlight your most important data or show data trends. Use the Quick Analysis tool for a Live Preview to try it out.

(1) Select the data you want to examine more closely.

(2) Click the Quick Analysis button in the bottom-right corner of the selection.

(3) Explore the options on the Formatting and Sparklines tabs to see how they affect your data(ref. Fig.2-4-10).

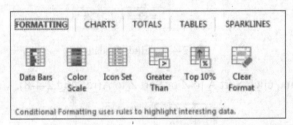

Fig.2-4-10　Formatting tabs

For example, pick a color scale in the Formatting gallery to differentiate high, medium, and low temperatures(ref. Fig.2-4-11).

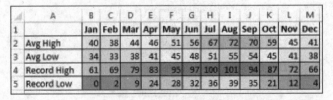

Fig.2-4-11　Color scale

(4) When you like what you see, click that option.

- **Show your data in a chart**

The Quick Analysis tool recommends the right chart for your data and gives you a visual presentation in just a few clicks.

(1) Select the cells that contain the data you want to show in a chart.

(2) Click the Quick Analysis button 🖻 in the bottom-right corner of the selection.

(3) Click the Charts tab, move across the recommended charts to see which one looks best for your data, and then click the one that you want(ref. Fig.2-4-12).

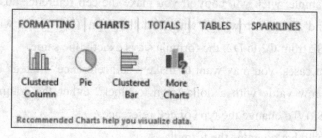

Fig.2-4-12   Charts tab

NOTE: Excel shows different charts in this gallery, depending on what's recommended for your data.

- **Save your work**

(1) Click the Save button on the Quick Access Toolbar, or press Ctrl+S(ref. Fig.2-4-13).

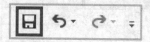

Fig.2-4-13   Save button

If you've saved your work before, you're done.

(2)   If this is the first time you've save this file:

(a) Under Save As, pick where to save your workbook, and then browse to a folder.

(b) In the File name box, enter a name for your workbook.

(c) Click Save.

- **Print your work**

(1) Click File, and then click Print, or press Ctrl+P.

(2) Preview the pages by clicking the Next Page and Previous Page arrows.

◀ 1 of 3 ▶

The preview window displays the pages in black and white or in color, depending on your printer settings.

If you don't like how your pages will be printed, you can change page margins or add page breaks.

(3) Click Print.

2. Switch between relative, absolute, and mixed references

By default, a cell reference is relative. For example, when you refer to cell A2 from cell C2, you are actually referring to a cell that is two columns to the left (C minus A), and in the same row (2). A formula that contains a relative cell reference changes as you copy it from one cell to another.

As an example, if you copy the formula=A2+B2 from cell C2 to D2, the formula in D2 adjusts to the right by one column and becomes=B2+C2. If you want to maintain the original cell reference in this example when you copy it, you make the cell reference absolute by preceding the columns (A and B) and row (2) with a dollar sign ($). Then, when you copy the formula=$A$2+$B$2 from C2 to D2, the formula stays exactly the same.

In less frequent cases, you may want to make a cell reference "mixed" by preceding either the column or the row value with a dollar sign to "lock" either the column or the row (for example, $A2 or B$3). To change the type of cell reference:

(1) Select the cell that contains the formula.

(2) In the formula bar [fx   ], select the reference that you want to change.

(3) Press F4 to switch between the reference types.

The following table (ref. Fig.2-4-14) summarizes how a reference type updates if a formula containing the reference is copied two cells down and two cells to the right(ref. Table 2-4-1).

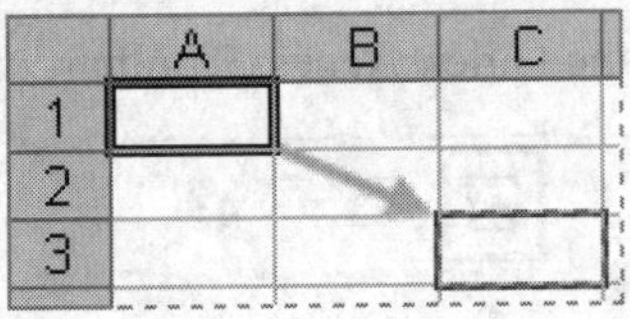

Fig.2-4-14　Cell

Table 2-4-1　Cell reference change

If the reference is	It changes to
$A$1 (absolute column and absolute row)	$A$1 (the reference is absolute)
A$1 (relative column and absolute row)	C$1 (the reference is mixed)
$A1 (absolute column and relative row)	$A3 (the reference is mixed)
A1 (relative column and relative row)	C3 (the reference is relative)

3. The parts of an Excel formula

A formula can also contain any or all of the following: functions, references, operators, and constants.

Parts of a formula:

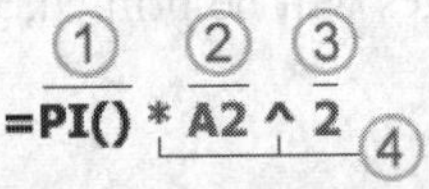

=PI() * A2 ^ 2
　①　　②　③
　　　　　　④

(1) Functions: The PI() function returns the value of pi: 3.142...

(2) References: A2 returns the value in cell A2.

(3) Constants: Numbers or text values entered directly into a formula, such as 2.

(4) Operators: The ^ (caret) operator raises a number to a power, and the * (asterisk) operator multiplies numbers.

4. Using calculation operators in Excel formulas

Operators specify the type of calculation that you want to perform on the elements of a formula. Excel follows general mathematical rules for calculations. Using parentheses allows you to change that calculation order.

Types of operators. There are four different types of calculation operators: arithmetic, comparison, text concatenation, and reference.

- ***Arithmetic operators***

To perform basic mathematical operations, such as addition, subtraction, multiplication, or division; combine numbers; and produce numeric results, use the following arithmetic operators(ref. Table 2-4-2).

Table 2-4-2   Arithmetic operator

Arithmetic operator	Meaning	Example
+ (plus sign)	Addition	=3+3
– (minus sign)	Subtraction Negation	=3–3
* (asterisk)	Multiplication	=3*3
/ (forward slash)	Division	=3/3
% (percent sign)	Percent	=30%
^ (caret)	Exponentiation	=3^3

- ***Comparison operators***

You can compare two values with the following operators. When two values are compared by using these operators, the result is a logical value—either TRUE or FALSE(ref. Table 2-4-3).

Table 2-4-3   Comparison operator

Comparison operator	Meaning	Example
= (equal sign)	Equal to	=A1=B1
> (greater than sign)	Greater than	=A1>B1
< (less than sign)	Less than	=A1<B1
>= (greater than or equal to sign)	Greater than or equal to	=A1>=B1
<= (less than or equal to sign)	Less than or equal to	=A1<=B1
<> (not equal to sign)	Not equal to	=A1<>B1

- ***Text concatenation operator***

Use the ampersand (&) to concatenate (join) one or more text strings to produce a single piece of text(ref. Table 2-4-4).

Table 2-4-4 Text operator

Text operator	Meaning	Example
& (ampersand)	Connects, or concatenates, two values to produce one continuous text value	="North"&"wind" results in "Northwind"

- *Reference operators*

Combine ranges of cells for calculations with the following operators(ref. Table 2-4-5).

Table 2-4-5 Reference operator

Reference operator	Meaning	Example
: (colon)	Range operator, which produces one reference to all the cells between two references, including the two references.	=B5:B15
, (comma)	Union operator, which combines multiple references into one reference	=SUM(B5:B15,D5:D15)
(space)	Intersection operator, which produces one reference to cells common to the two references	=B7:D7 C6:C8

5. Overview of formulas in Excel

If you're new to Excel, you'll soon find that it's more than just a grid in which you enter numbers in columns or rows. Sure, you can use Excel to find totals for a column or row of numbers, but you can also calculate a mortgage payment, solve math or engineering problems, or find a best case scenario based on variable numbers that you plug in.

Excel does this by using formulas in cells. A formula performs calculations or other actions on the data in your worksheet. A formula always starts with an equal sign (=), which can be followed by numbers, math operators (like a + or - sign for addition or subtraction), and built-in Excel functions, which can really expand the power of a formula.

For example, the following formula multiplies 2 by 3 and then adds 5 to that result to come up with the answer, 11.

=2*3+5

Here are some additional examples of formulas that you can enter in a worksheet.

- =A1+A2+A3   Adds the values in cells A1, A2, and A3.
- =SUM(A1:A10)   Uses the SUM function to return sum of the values in A1 through A10.
- =TODAY()   Returns the current date.
- =UPPER("hello")   Converts the text "hello" to "HELLO" by using the UPPER function.
- =IF(A1>0)   Uses the IF function to test the cell A1 to determine if it contains a value greater than 0.

Unit 5 Word Training

Word 2016 is designed to help you create professional-quality documents. Word can also help you organize and write documents more efficiently.

When you create a document in Word, you can choose to start from a blank document or let a template do much of the work for you. From then on, the basic steps in creating and sharing documents are the same. And Word's powerful editing and reviewing tools can help you work with others to make your document great.

TIP: To learn about new features, see What's new in Word 2016.

- **Start a document**

It's often easier to create a new document using a template instead of starting with a blank page. Word templates come ready-to-use with pre-set themes and styles. All you need to do is add your content.

Each time you start Word, you can choose a template from the gallery, click a category to see more templates, or search for more templates online.

For a closer look at any template, click it to open a large preview.

If you'd rather not use a template, click Blank document (ref. Fig.2-5-1).

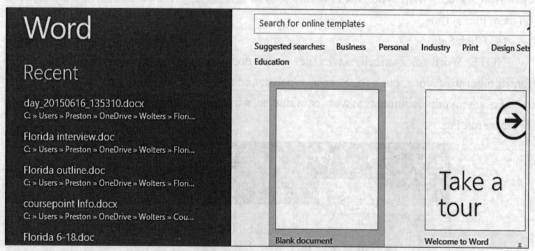

Fig.2-5-1 Start a document

- **Open a document**

Every time you start Word, you'll see a list of your most recently used documents in the left column. If the document you're looking for isn't there, click Open Other Documents (ref.

Fig.2-5-2).

Fig.2-5-2 Open a document

If you're already in Word, click File→Open and then browse to the file's location.

When you open a document that was created in earlier versions of Word, you see Compatibility Mode in the title bar of the document window. You can work in compatibility more or you can upgrade the document to use Word 2016. To learn more, see Use Word 2016 to open documents created in earlier versions of Word.

● **Save a document**

To save a document for the first time, do the following:

(1) On the File tab, click Save As.

(2) Browse to the location where you'd like to save your document.

NOTE: To save the document on your computer, choose a folder under This PC or click Browse. To save your document online, choose an online location under Save As or click Add a Place. When your files are online, you can share, give feedback and work together on them in real time.

(3) Click Save.

NOTE: Word automatically saves files in the .docx file format. To save your document in a format other than .docx, click the Save as typelist, and then select the file format that you want.

To save your document as you continue to work on it, click Save in the Quick Access Toolbar (ref. Fig.2-5-3).

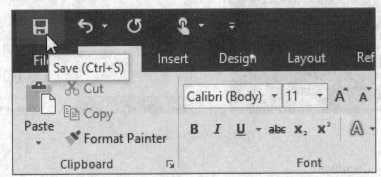

Fig.2-5-3 Save a document

- **Read documents**

Open your document in Read Mode to hide most of the buttons and tools so you can get absorbed in your reading without distractions (ref. Fig.2-5-4).

Fig.2-5-4　Read documents

(1) Open the document you want to read.

NOTE: Some documents open in Read Mode automatically, such as protected documents or attachments.

(2) Click View→Read Mode.

(3) To move from page to page in a document, do one of the following:

- Click the arrows on the left and right sides of the pages.
- Press page down and page up or the spacebar and backspace on the keyboard. You can also use the arrow keys or the scroll wheel on your mouse.
- If you're on a touch device, swipe left or right with your finger.

TIP: Click View→Edit Document to edit the document again.

- **Track changes**

When you're working on a document with other people or editing a document yourself, turn on Track Changes to see every change. Word marks all additions, deletions, moves, and formatting changes.

(1) Open the document to be reviewed.

(2) Click Review and then on the Track Changes button, select Track Changes (ref. Fig.2-5-5).

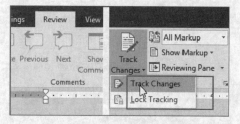

Fig.2-5-5　Track Changes

Read Track changes or Remove tracked changes and comments to learn more.

● **Print your document**

All in one place, you can see how your document will look when printed, set your print options, and print the file.

(1) On the File tab, click Print (ref. Fig.2-5-6).

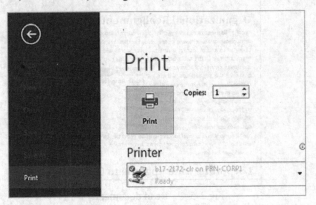

Fig.2-5-6　Print your document

(2) Do the following:
- Under Print, in the Copies box, enter the number of copies you want.
- Under Printer, make sure the printer you want is selected.
- Under Settings, the default print settings for your printer are selected for you. If you want to change a setting, just click the setting you want to change and then select a new setting.

(3) When you're satisfied with the settings, click Print.

Unit 6　PowerPoint Training

Basic tasks for creating a PowerPoint presentation

Applies To: PowerPoint 2016 PowerPoint 2013

PowerPoint presentations work like slide shows. To convey a message or a story, you break it down into slides. Think of each slide as a blank canvas for the pictures, words, and shapes that will help you build your story.

NOTE: For information about earlier versions of PowerPoint, see Create a basic presentation in PowerPoint.

- **Choose a theme**

When you open PowerPoint, you'll see some built-in themes and templates. A theme is a slide design that contains matching colors, fonts, and special effects like shadows, reflections, and more.

(1) Choose a theme.

(2) Click Create, or pick a color variation and then click Create (ref. Fig.2-6-1).

Fig.2-6-1　Choose a theme

Read more: Apply color and design to my slides (theme)

- **Insert a new slide**

On the Home tab, click New Slide, and pick a slide layout (ref. Fig.2-6-2).

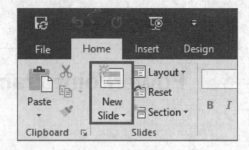

Fig.2-6-2　New Slide

- **Save your presentation**

(1) On the File tab, choose Save.
(2) Pick or browse to a folder.
(3) In the File name box, type a name for your presentation, and then choose Save.

NOTE: If you frequently save files to a certain folder, you can 'pin' the path so that it is always available (as shown below) (ref. Fig.2-6-3).

Fig.2-6-3　Save file

TIP: Save your work as you go. Hit Ctrl+S often.

- **Add text**

Select a text placeholder, and begin typing (ref. Fig.2-6-4).

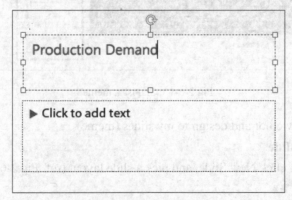

Fig.2-6-4　Add text

- **Format your text**

(1) Select the text.

(2) Under Drawing Tools, choose Format (ref. Fig.2-6-5).

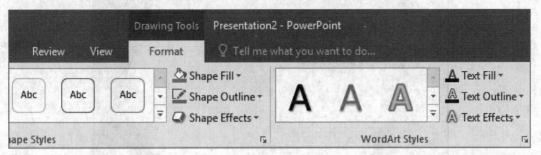

Fig.2-6-5   Choose Format

(3) Do one of the following:

To change the color of your text, choose Text Fill, and then choose a color.

To change the outline color of your text, choose Text Outline, and then choose a color.

To apply a shadow, reflection, glow, bevel, 3-D rotation, a transform, choose Text Effects, and then choose the effect you want.

- **Add pictures**

On the Insert tab, do one of the following:

(1) To insert a picture that is saved on your local drive or an internal server, choose Pictures, browse for the picture, and then choose Insert.

(2) To insert a picture from the Web, choose Online Pictures, and use the search box to find a picture. (ref. Fig.2-6-6)

(3) Choose a picture, and then click Insert.

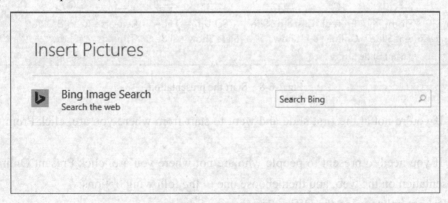

Fig.2-6-6   Insert a picture

- **Add speaker notes**

Slides are best when you don't cram in too much information. You can put helpful facts and notes in the speaker notes, and refer to them as you present.

(1) To open the notes pane, at the bottom of the window, click Notes ≜ Notes .

(2) Click inside the Notes pane below the slide, and begin typing your notes (ref. Fig.2-6-7).

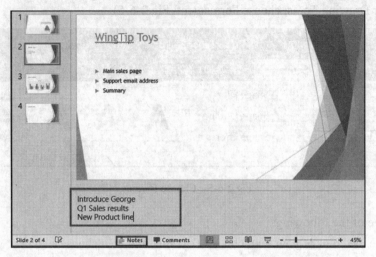

Fig.2-6-7　Add speaker notes

● **Give your presentation**

On the Slide Show tab, do one of the following:

(1) To start the presentation at the first slide, in the Start Slide Show group, click From Beginning. (ref. Fig.2-6-8)

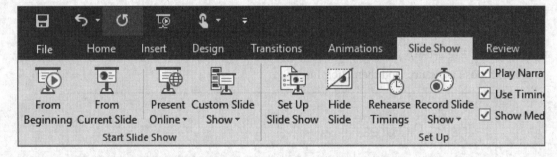

Fig.2-6-8　Start the presentation

(2) If you're not at the first slide and want to start from where you are, click From Current Slide.

(3) If you need to present to people who are not where you are, click Present Online to set up a presentation on the web, and then choose one of the following options:

- Present online using the Office Presentation Service.
- Start an online presentation in PowerPoint using Skype for Business.

Unit 7　Access Training

1. Basic tasks for an Access desktop database

Applies To: Access 2016 Access 2013

Access desktop databases can help you store and track just about any kind of information, such as inventory, contacts, or business processes. Let's take a walk through the paths you can take to create an Access desktop database, add data to it, and then learn about next steps towards customizing and using your new database.

- **Choose a template**

Access templates have built-in tables, queries, forms, and reports that are ready to use. A choice of templates is the first thing you'll notice when you start Access, and you can search online for more templates. (ref. Fig.2-7-1)

Fig.2-7-1　Access templates

(1) In Access click File→New.

(2) Select a desktop database template and enter a name for your database under File Name. (If you don't see a template that would work for you, use the Search online templatesbox.)

(3) You can either use the default location that Access shows below the File Name box or click the folder icon to pick one.

(4) Click Create (ref. Fig.2-7-2).

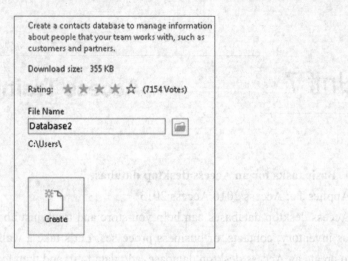

Fig.2-7-2　New File

Depending on the template, you might need to do any of the following to get started:

(1) If Access displays a Login dialog box with an empty list of users:

a. Click New User.

b. Fill in the User Details form.

c. Click Save & Close.

d. Select the user name you just entered, and then click Login.

(2) If Access displays a Security Warning message in the message bar, and you trust the source of the template, click Enable Content. If the database requires a login, log in again.

For more, see create an Access desktop database from a template.

● **Create a database from scratch**

If none of the templates fit your needs, you might start with a blank desktop database.

(1) From Access, click New→Blank desktop database.

(2) Type a name for your database in the File Name box.

(3) You can either use the default location that Access shows below the File Name box or click the folder icon to pick one.

(4) Click Create.

● **Add a table**

In a database, your information is stored in multiple related tables. To create a table:

(1) When you open your database for the first time, you'll see a blank table in Datasheet view where you can add data. To add another table, click the Create tab→Table. You can either start entering data in the empty field (cell) or paste data from another source like an Excel workbook.

(2) To rename a column (field), double-click the column heading, and then type the new name.

TIP: Meaningful names help you know what each field contains without seeing its contents.

(3) Click File→Save.

- To add more fields, type in the Click to Add column.
- To move a column, select it by clicking its column heading, and then drag it to where you want it. You can also select contiguous columns and drag them all to a new location. For more, see Introduction to tables.

● **Copy and paste data**

You can copy and paste data from another program like Excel or Word into an Access table. This works best if the data is separated into columns. If the data is in a word processing program, such as Word, either use tags to separate the columns or convert into a table format before copying.

(1) If the data needs editing, such as separating full names into first and last names, do that first in the source program.

(2) Open the source and copy (Ctrl + C) the data.

(3) Open the Access table where you want to add the data in Datasheet view and paste it (Ctrl + V).

(4) Double-click each column heading and type a meaningful name.

(5) Click File→Save and give your new table a name.

NOTE: Access sets the data type of each field based on the information you paste into the first row of each column, so make sure that the information in the following rows match the first row.

● **Import or link to data**

You can either import data from other sources, or you can link to the data from Access without moving the information from where it is stored. Linking can be a good option if you have multiple users updating the data and you want to make sure that you are seeing the latest version or if you want to save storage space. You can choose whether you want to link to or import data for most formats. (ref. Fig.2-7-3)

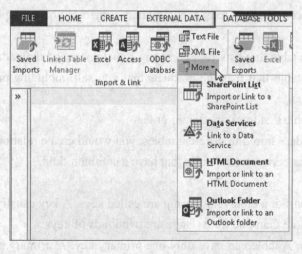

Fig.2-7-3　Import or link to data

The process differs slightly depending on the data source, but these instructions will get you started:

(1) On the External Data tab, click the data format you'll be importing from or linking to. If you don't see the right format, click More.

(2) Follow the instructions in the Get External Data dialog box.

2. Introduction to tables

● **Overview**

A relational database like Access usually has several related tables. In a well-designed database, each table stores data about a particular subject, such as employees or products. A table has records (rows) and fields (columns). Fields have different types of data, such as text, numbers, dates, and hyperlinks. (ref. Fig.2-7-4)

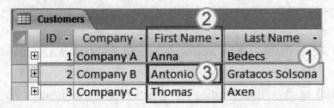

Fig.2-7-4   Table

(1) A record: Contains specific data, like information about a particular employee or a product.

(2) A field: Contains data about one aspect of the table subject, such as first name or e-mail address.

(3) A field value: Each record has a field value. For example, Contoso, Ltd. or someone@example.com.

● **Table relationships**

Although each table in a database stores data about a specific subject, tables in a relational database such as Access, store data about related subjects. For example, a database might contain:

(1) A customers table that lists your company's customers and their addresses.

(2) A products table that lists the products that you sell, including prices and pictures for each item.

(3) An orders table that tracks customer orders.

To connect the data stored in different tables, you would create relationships. A relationship is a logical connection between two tables that have a common field.

● **Keys**

Fields that are part of a table relationship are called keys. A key usually consists of one field, but may consist of more than one field. There are two kinds of keys:

(1) Primary key: A table can have only one primary key. A primary key consists of one or more fields that uniquely identify each record that you store in the table. Access automatically

provides a unique identification number, called an ID number that serves as a primary key.

(2) Foreign key: A table can have one or more foreign keys. A foreign key contains values that correspond to values in the primary key of another table. For example, you might have an Orders table in which each order has a customer ID number that corresponds to a record in a Customers table. The customer ID field is a foreign key of the Orders table.

The correspondence of values between key fields forms the basis of a table relationship. You use a table relationship to combine data from related tables. For example, suppose that you have a Customers table and an Orders table. In your Customers table, each record is identified by the primary key field, ID.

To associate each order with a customer, you add a foreign key field to the Orders table that corresponds to the ID field of the Customers table, and then create a relationship between the two keys. When you add a record to the Orders table, you use a value for customer ID that comes from the Customers table. Whenever you want to view any information about an order's customer, you use the relationship to identify which data from the Customers table corresponds to which records in the Orders table (ref. Fig.2-7-5).

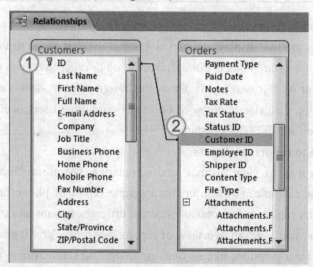

Fig.2-7-5   Table relationship

(1) A primary key is identified by the key icon next to the field name.

(2) A foreign key — note the absence of the key icon.

3. Introduction to queries

Using a query makes it easier to view, add, delete, or change data in your Access database. Some other reasons for using queries:

- Find specific quickly data by filtering on specific criteria (conditions).
- Calculate or summarize data.
- Automate data management tasks, such as reviewing the most current data on a recurring basis.

You get a more robust set of query options when you work with an Access desktop database, but Access web apps offer some of the query options shown below.

- **Queries help you find and work with your data**

In a well-designed database, the data that you want to present through a form or report is usually located in multiple tables. A query can pull the information from various tables and assemble it for display in the form or report. A query can either be a request for data results from your database or for action on the data, or for both. A query can give you an answer to a simple question, perform calculations, combine data from different tables, add, change, or delete data from a database. Since queries are so versatile, there are many types of queries and you would create a type of query based on the task (ref. Table 2-7-1).

Table 2-7-1 Major query types

Major query types	Use
Select	To retrieve data from a table or make calculations
Action	Add, change, or delete data. Each task has a specific type of action query. Action queries are not available in Access apps

4. Create a form in Access

Forms in Access are like display cases in stores that make it easier to view or get the items that you want. Since forms are objects through which you or other users can add, edit, or display the data stored in your Access desktop database, the design of your form is an important aspect. If your Access desktop database is going to be used by multiple users, well-designed forms is essential for efficiency and data entry accuracy.

5. Introduction to reports in Access

Reports offer a way to view, format, and summarize the information in your Microsoft Access database. For example, you can create a simple report of phone numbers for all your contacts, or a summary report on the total sales across different regions and time periods.

From this article, you'll get an overview of reports in Access. You'll also learn the basics of creating a report, and using options like sorting, grouping, and summarizing the data, and how to preview and print the report.

- **Overview of reports in Access**

What can you do with a report?

A report is a database object that comes in handy when you want to present the information in your database for any of the following uses:

- Display or distribute a summary of data.
- Archive snapshots of the data.
- Provide details about individual records.
- Create labels.

- **Parts of a report**

While it is possible to create "unbound" reports that do not display data, but for the purposes of this article, we'll assume that a report is bound to a data source such as a table or query. The design of a report is divided into sections that you can view in the Design view. Understanding how each section works can helps you create better reports (ref. Table 2-7-2).

Table 2-7-2 Parts of a report

Section	How the section is displayed when printed	Where the section can be used
Report Header	At the beginning of the report	Use the report header for information that might normally appear on a cover page, such as a logo, a title, or a date. When you place a calculated control that uses the Sum aggregate function in the report header, the sum calculated is for the entire report. The report header is printed before the page header
Page Header	At the top of every page	Use a page header to repeat the report title on every page
Group Header	At the beginning of each new group of records	Use the group header to print the group name. For example, in a report that is grouped by product, use the group header to print the product name. When you place a calculated control that uses the Sum aggregate function in the group header, the sum is for the current group. You can have multiple group header sections on a report, depending on how many grouping levels you have added. For more information about creating group headers and footers, see the section Add grouping, sorting, or totals
Detail	Appears once for every row in the record source	This is where you place the controls that make up the main body of the report
Group Footer	At the end of each group of records	Use a group footer to print summary information for a group. You can have multiple group footer sections on a report, depending on how many grouping levels you have added
Page Footer	At the end of every page	Use a page footer to print page numbers or per-page information
Report Footer	At the end of the report	Use the report footer to print report totals or other summary information for the entire report. NOTE: In Design view, the report footer appears below the page footer. However, in all other views, the report footer appearsabove the page footer, just after the last group footer or detail line on the final page

附 录

附录一 习题参考答案

Unit 1   Hardware

Ex1. Multiple choice.
1. B   2. D   3. A   4. B   5. B   6. D   7. D   8. D   9. C
10. (1) A   (2) D   (3) C   (4) E   (5) F   (6) G   (7) H   (8) B   (9) I
11. (1) D   (2) E   (3) C   (4) B   (5) A

Ex2. Match each numbered item with the most closely related lettered item.
1. c   2. a   3. e   4. i   5. h   6. g   7. j   8. b   9. d   10. f

Ex3. Computer English test.
1. Answers: (1) B   (2) C   (3) A   (4) B   (5) B
2. Answers: (1) A   (2) B   (3) C   (4) D   (5) B
3. Answers: (1) A   (2) B   (3) D   (4) C

Ex4. Translate the following sentences into English.

1. The equipment that processes the data to create information is called hardware. It includes keyboard, mouse, monitor, system unit, and other devices. Hardware is controlled by software.

2. Computer is electronic device that can follow instructions to accept input, process input and produce information.

3. Microcomputers are the least powerful, yet are the most widely used and fastest-growing type of computer. Categories of microcomputer include: desktop, notebook and PDA.

4. Input devices translate data and program that humans can understand into a form that the computer can process.

5. Unlike memory, secondary storage devices hold data and program even after electrical power to the computer system has been turned off.

Ex5. Speaking. (略)

Unit 2   Software

Ex1. Multiple choice.
1. A   2. D   3. C   4. B   5. B   6. A   7. D   8. C   9. A   10. D   11. B

Ex2. Match each numbered item with the most closely related lettered item.

1. b   2. a   3. i   4. d   5. c   6. j   7. f   8. h   9. e   10. g

Ex3. Computer English test.

1. Answers: (1) D   (2) A   (3) B   (4) D   (5) B
2. Answers: (1) A   (2) B   (3) D   (4) D
3. Answers: (1) E   (2) D   (3) G   (4) A   (5) B

Ex4. Translate the following sentences into English.

1. Application software can be divided into two categories: general-purpose applications and special-purpose applications. General-purpose applications include browser, word processors, spreadsheets, database management systems and presentation graphics. Special-purpose applications include multimedia, graphics, virtual reality and artificial intelligence programs.

2. In order to connect to other resources with a browser, the location or address must be specified. These addressed are called URL. Once browser has connected to a web site, a document file is sent to your computer. Browser interprets the HTML commands contained in the document and displays the document as a web page. Typically, the first page of a web site is referred to as its home page.

3. Many operating system and application program use windows to display information or request input. More than one windows can be opened or displayed on the computer screen at one time. Windows can generally be resized, moved, and closed.

4. Data and information are stored in workbook, which can contain one or more related worksheets. The column are identified by letters and rows are identified by numbers. The intersection of a row and column creates a cell. Information is entered into cells.

5. Database is a collection of related data. DBMS are used to create and use database. A relational database organize data into related tables. Each table is made up of row called records and columns called fields.

6. Integrated package is a single program providing the functionality of a word processor, spreadsheet, database manager and more.

7. Word processor allow you to create, edit, save and print text-based document such as letters and report.

Ex5. Translate the following sentences into Chinese.

1. 文件未找到。
2. 错误命令或文件名。
3. 无效驱动器标志。
4. 文件不能复制到自身。
5. 无效路径、非目录或目录非空
6. 删除一个或多个文件。
7. 创建一个目录。
8. 重命名一个或多个文件。
9. 显示或设置系统时间。

10. 显示文本文件的内容。

Ex6. Speaking. (略)

Unit 3   Operating Systems

Ex1. Multiple choice.

1. A   2. B   3. D   4. B

5. (1) B   (2) A   (3) D   (4) F   (5) D

6. (1) F   (2) B   (3) G   (4) C   (5) D   (6) E   (7) A

Ex2. Match each numbered item with the most closely related lettered item.

1. h   2. b   3. d   4. f   5. g   6. c   7. i   8. j   9. a   10. e

Ex3. Computer English test.

1. Answers: (1) C   (2) A   (3) B   (4) D

2. Answers: (1) I   (2) E   (3) C   (4) H   (5) A

Ex4. Translate the following sentences into English.

1. Language translator convert the programming instructions written by Programmers into a language that computer understand and process.

2. System software consists of operating system, utilities, device driver and language translator. Operating system perform three basic functions: manage resources, provide user interface and run application software.

3. The desktop is the user interface provided by Windows. Windows store information in files and folders. Icons are often used to interact with the Windows operating system. Another powerful, easy to use operating system designed to run on Macintosh computers is Mac OS.

4. Device drivers are specialized programs that work with operating system to allow communication between hardware devices and the computer system.

Ex5. Speaking. (略)

Unit 4   Data

Ex1. Multiple choice.

1. A   2. D   3. B   4. C   5. A   6. D

7. Answers: (1) A   (2) C   (3) D   (4) E   (5) F   (6) B

Ex2. Match each numbered item with the most closely related lettered item.

1. c   2. e   3. g   4. b   5. d   6. f   7. a

Ex3. Computer English test.

1. Answers: (1) D   (2) C   (3) B   (4) A   (5) C

2. Answers: (1) A   (2) B   (3) B   (4) C   (5) C

Ex4. Translate the following sentences into English.

1. Internet is the world's largest computer network. Web is an Internet service that provide a

multimedia interface to resources available on the Internet.

2. Microcomputer hardware consists of a variety of different devices. The three categories of devices are: system unit, input/output devices and secondary storage. secondary storage device store data and programs. Typical media include floppy, hard and optical disks.

3. Data is typically stored electronically in a file. Common types of file are: document file, worksheet file, database file and presentation file.

4. There are two major types of software: system software and application software. The user interact with application software. System software enables the application software to interact with the computer hardware.

5. Software is a another name of program. Program consists of step-by-step instructions that tell the computer how to do its work. The purpose of software is to convert data into information.

Ex5. Speaking. (略)

Unit 5　Programming

Ex1. Multiple choice.
1. A　2. C　3. C　4. B　5. D　6. C　7. C　8. C　9. B　10. B

Ex2. Match each numbered item with the most closely related lettered item.
1. i　2. h　3. j　4. a　5. b　6. d　7. e　8. g　9. c　10. f

Ex3. Computer English test.
1. Answers: (1) A　(2) B　(3) A　(4) C
2. Answers: (1) A　(2) B　(3) D　(4) C　(5) C

Ex4. Translate the following sentences into English.

1. Backup programs make copies of files to be used in case the originals are lost or damaged.

Windows operating systems are accompanied with several utilities including backup, disk cleanup and disk defragmenter.

2. Cache acts as a temporary high-speed holding area between memory and CPU.

Cache memory is used to store the most frequently accessed information stored in RAM.

3. Arithmetic and logic unit, usually called ALU perform two types of operations—arithmetic and logical. Arithmetic operations are the functional math operations: addition, subtraction, multiplication and division. Logical operations consist of comparison.

4. RAM chips are called temporary or volatile because the contents are lost if power is disrupted.

Ex5. Translate the following sentences into Chinese.
1. 文件未找到。
2. 错误命令或文件名。
3. 无效驱动器标识。
4. 文件不能复制到自身。

5. 无效路径、非目录或目录非空。

Ex6. Speaking. (略)

Unit 6　Computer Science

Ex1. Multiple choice.

Answers: (1) F　　(2) C　　(3) E　　(4) J　　(5) A

Ex2. Match each numbered item with the most closely related lettered item.

1. f　　2. d　　3. e　　4. c　　5. a　　6. b

Ex3. Computer English test.

1. Answers: (1) D　　(2) F　　(3) A　　(4) H　　(5) H
2. Answers: (1) C　　(2) E　　(3) G　　(4) F　　(5) A
3. Answers: (1) A　　(2) A　　(3) B　　(4) A　　(5) D

Ex4. Translate the following sentences into English.

1. Terminal is an input and output device that connect you to a mainframe or other type of computer called a host or server.

2. The most common point device is the mouse. Three basic mouse design include mechanical mouse, optical mouse, and wireless mouse. Devices similar to mouse include trackballs, touch screen, and pointing sticks.

3. The images displayed on the monitor are called soft copy, the information output on the paper is called hard copy. There are several types of printers: ink-jet printer, laser printer, thermal printer and other printers including dot-matrix, chain printer and plotters.

Ex5. Speaking. (略)

Unit 7　Multimedia

Ex1. Multiple choice.

1. C　2. A　3. D　4. C　5. B　6. B　7. D　8. A
9. Answers: (1) B　　(2) D　　(3) C　　(4) E　　(5) A

Ex2. Match each numbered item with the most closely related lettered item.

1. f　2. a　3. i　4. b　5. j　6. e　7. d　8. h　9. c　10. g

Ex3. Computer English test.

1. Answers: (1) B　　(2) B　　(3) A　　(4) D　　(5) C
2. Answers: (1) B　　(2) C　　(3) D　　(4) A　　(5) B
3. Answers: (1) D　　(2) F　　(3) E　　(4) E　　(5) C
4. Answers: (1) B　　(2) C　　(3) D　　(4) E　　(5) A

Ex4. Translate the following sentences into English.

1. The most common ways to input data is by keyboard. Keyboards convert numbers, letters and special characters that people understand into electronic signals.

2. Digital cameras are similar to traditional cameras except that images are recorded digitally on a disk or in the camera's memory rather than on film.

3. Audio input device convert sounds into a form that can be processed by a computer. These sounds can be from a wide variety of sources. By far the most widely used audio input device is the microphone.

4. The most frequently used output device is monitor. Two important characterristics of monitor are size and clarity. A monitor's clarity is indicated by its resolution, which is measured in pixels. For a given monitor, the greater the resolution, the better the clarity of the image.

Ex5. Speaking. (略)

Unit 8   Networks

Ex1. Multiple choice.
1. D   2. C   3. E   4. C   5. A   6. D   7. B   8. C   9. C   10. A
11. (1) B   (2) C   (3) E   (4) D   (5) A
12. (1) D   (2) E   (3) B   (4) C   (5) A

Ex2. Match each numbered item with the most closely related lettered item.
1. c   2. a   3. g   4. b   5. l   6. d   7. e   8. i   9. j   10. f
11. k   12. h

Ex3. Computer English test.
1. Answers: (1) C   (2) C   (3) B   (4) A   (5) C
2. Answers: (1) C   (2) C   (3) A   (4) B   (5) D

Ex4. Translate the following sentences into English.

1. Sever shares resources with other nodes. Depending on the resources shard, it may be called a file server, print server, communication server, Web server or database server.

2. Physical connections use a solid medium such as twisted pair telephone line, coaxial cable, and fiber-optic cable. Wireless connection use air rather than solid substance to connect sending and receiving devices. Two primary technologies are microwave and satellite.

3. In a bus network, each device in the network handles its own communications control. There is no host computer. All communications travel along a common connecting cable called a bus. As the information passes along the bus, it is examined to see if the information is intended for it .

Ex5. Speaking. (略)

Unit 9   Internet and Online Services

Ex1. Multiple choice.
1. A   2. A   3. D   4. C   5. D   6. B   7. B   8. (1) C   (2) B   9. B   10. C
11. (1) C   (2) D   (3) B   (4) E   (5) A

12. (1) E   (2) C   (3) D   (4) A   (5) B

Ex2. Match each numbered item with the most closely related lettered item.

1. d   2. h   3. a   4. i   5. b   6. k   7. e   8. l   9. g   10. f   11. c   12. j

Ex3. Computer English test.

1. Answers: (1) B   (2) C   (3) A   (4) D   (5) B
2. Answers: (1) C   (2) A   (3) D   (4) D   (5) B

Ex4. Translate the following sentences into English.

1. Floppy disks are a portable storage media. They are used to store and to transport word processing, spreadsheet, and other types of files.

2. File compression and file decompression increase storage capacity by reducing the amount of space required to store data and programs. File compression is not limited to hard disk systems. It is frequently used to compress files on floppy disks as well.

3. Flash memory is an example of solid state storage. It is typically used to store digitized images and record MP3 files.

4. Magnetic tape provides sequential access, disks provides direct access. Magnetic tape is primarily used for backing up data.

Ex5. Speaking. (略)

Unit 10　　World Wide Web

Ex1. Multiple choice.

1. D   2. B   3. A   4. B   5. B   6. B
7. (1) B   (2) D   (3) E   (4) C   (5) A
8. (1) C   (2) D   (3) E   (4) B   (5) A
9. (1) B   (2) D   (3) C   (4) E   (5) A

Ex2. Match each numbered item with the most closely related lettered item.

1. c   2. e   3. i   4. a   5. d   6. g   7. h   8. b   9. j   10. f

Ex3. Computer English test.

1. Answers:(1) C   (2) B   (3) A   (4) B   (5) C
2. Answers:(1) B   (2) D   (3) A   (4) D   (5) C

Ex4. Translate the following sentences into English.

1. Plug-ins are automatically loaded and operate as part of a browser. Some are included in many today's browsers. Other must be installed.

2. Intranet are private networks within an organization that resemble intranet. They use browsers, web pages that are available only to those within the organization.

3. Firewall is a security system to protect against external threats. It consists of both Hardware and software. All communication into and out of an organization pass through a special security computer called proxy server.

Ex5. Speaking. (略)

附录二　专业英语样卷

试　卷　一

Ⅰ．Translate the following key words into English. (20%)

(1) 电子邮件地址　　　　　　　　(2) 图像分辨率
(3) 基本输入/输出系统　　　　　　(4) 双击
(5) 无连接　　　　　　　　　　　(6) 扩展内存区
(7) 超文本　　　　　　　　　　　(8) 协议端口
(9) 存储设备　　　　　　　　　　(10) 颜色深度

Ⅱ．Translate the following key words into Chinese. (20%)

(1) Image projection technology　　(2) Cathode ray tube
(3) Flat-panel displays　　　　　　(4) Mouse and modem
(5) Physical structure of the net　　(6) Distributed hypermedia
(7) Domain name system　　　　　(8) High-level language
(9) Parallel operation　　　　　　(10) Connection-oriented

Ⅲ．Multiple choice. (30%)

1. The software that translates computer domain names into __(1)__ Internet addresses provides an interesting example of client-server interaction. The database of names is not kept on a single computer. Instead, the naming information is __(2)__ among a potentially large set of servers located at sites __(3)__ the Internet. Whenever an application program needs to translates a name, the application becomes a __(4)__ of the naming system. The client sends a request message to a name server, which finds the corresponding address and sends a reply message. If it cannot answer a request, a name server __(5)__ becomes the client of another name server, until a server is found that can answer the request.

(1)~(5)：A. client　　B. temporarily　　C. kept　　D. across　　E. equivalent

Answers: (1)_____ (2)_____ (3)_____ (4)_____ (5)_____

2. Future Computer Trends.

The components of a PC will be built into the __(1)__ and a large plasma or electroluminescent flat panel display hanging on a partition wall will at as both a __(2)__ and video phone display.

A high resolution, tough sensitive LCD __(3)__ into desktop will allow penbased pointing and data input. In the home, meanwhile, computers will become as ubiquitous as TVs.

Computer interfaces will be far friendlier than __(4)__ graphical user interfaces (GUI) of today. Computers will use human interfaces based on multimedia.

RISC microprocessors will finally obliterate the ever thinning gap between workstations and PCs. Similar advances in nonvolatile flash memory will lead to the replacement of magnetic __(5)__ Disks with electronic "silicon disks".

(1)~(5) A. build   B. monitor   C. desk   D. built   E. mouse driven
           F. keyboard   G. touch sensitive   H. soft   I. hard   J. optical

Answers: (1)_____ (2)_____ (3)_____ (4)_____ (5)_____

3. A computer system consists of several basic __(1)__. An input device provides data. The data are stored in __(2)__, which also holds a program. Under control of that program the computer's __(3)__ manipulates the data, storing the results back into memory. Finally, the results flow from the computer to an __(4)__ device. Additionally, most modern computers use secondary storage to extend memory __(5)__.

(1) A. parts   B. components   C. ingredient   D. assembly
(2) A. floppy disk   B. hard disk   C. memory   D. tape
(3) A. processor   B. heart   C. controller   D. plate
(4) A. input   B. output   C. import   D. export
(5) A. capability   B. capacity   C. content   D. size

Answers: (1)_____ (2)_____ (3)_____ (4)_____ (5)_____

Ⅳ. Translate the following four paragraphs into Chinese. (30%)

1. In printf statement, it is extremely important that the number of operators in the format string corresponds exactly with the number and type of the variables following it. For example, if the format string contains three operators and it must be followed by exactly three parameters, and they must have the same types in the same order as those specified by the operators.

2. On a time-shared system, users access a computer through terminals. Some terminals are local, linked directly to a computer by cables, while others are remote, communicating with distant computers over telephone lines or other transmission media. Computers, however, do not store data as continuous waves; they store and manipulate discrete pulses(脉冲). Because of this electronic incompatibility, whenever data are transmitted between a computer and a remote terminal, they must be converted from pulse form to wave form, and back again. Converting to wave form is called modulation; converting back is called demodulation; the task is performed by a hardware device called modem. Normally, there is one at each end of a communication line.

3. Microsoft MS-DOS contains several memory utilities to customize the way your personal computer uses memory. Be ware, however, that the total amount of memory in your computer is fixed (unless you purchase additional memory); and that if you configure a portion of memory using one utility, less of it remains to be used in other ways. For example, if you configure a portion of memory as expanded memory, less total memory remains to be used as extended memory. Similarly, if you devote a large amount of memory to a disk cache, less memory is

available for either extended or expanded memory.

试卷一参考答案

Ⅰ．关键词翻译（汉译英）。(20%)

(1) email address　(2) image resolution　(3) BIOS(Basic Input Output System)
(4) double click　(5) connectionless　(6) extended memory area　(7) hypertext
(8) protocol port　(9) storage device　(10) color depth

Ⅱ．关键词翻译（英译汉）。(20%)

(1) 图像投影技术　(2) 阴极射线管　(3)平面显示器　(4)鼠标与调制解调器　(5)网络物理结构　(6) 分布式超媒体　(7) 域名系统　(8) 高级语言　(9) 并行操作　(10) 面向连接的

Ⅲ．选择填空。(30%)

1. (1) E　(2) C　(3) D　(4) A　(5) B
2. (1) C　(2) B　(3) D　(4) E　(5) I
3. (1) B　(2) C　(3) A　(4) B　(5) B

Ⅳ．英译汉。(30%)

1. 必须注意，在 printf 语句中，格式字符串中的操作符的个数必须与后续的变量的个数和类型一致。例如，如果格式字符串中包含了 3 个操作符，则它后面必须有 3 个参数，并且参数在出现的顺序和类型上要与前面操作符规定的一致。

2. 在分时系统上，用户通过终端访问计算机。一些终端是本地的，用电缆直接连到计算机上，而另一些是远程的，通过电话线或其他传输介质与远程计算机通信。然而计算机并不按连续的波形存储数据，而是存储和处理离散脉冲。由于这种电子不兼容性，任何时候在计算机与远程终端之间传送数据，数据都必须从脉冲形式转换成波形，再从波形转换成脉冲形式。将脉冲形式转换成波形称为调制，将波形转换回脉冲形式称为解调。该任务由称作调制解调器的硬件完成。通常，在通信线路的两端都有一个调制解调器。

3. 微软的 MS-DOS 包含几个内存实用程序，用于定制 PC 内存的使用方式。然而要注意，你的计算机的内存总量是固定的（除非你购买附加的内存）；如果你配置了一部分使用某种实用程序的内存（将某些内存分配给一个内存管理程序使用后），可供其他方式使用的内存就相应减少了。同理，如果你将大量内存用作磁盘缓冲区，可用作扩展或扩充内存的内存空间就减少了。

试　卷　二

Ⅰ．Translate the following key words into English. (20%)

(1) 域名服务器　　　　　　　　　(2) 程序设计语言
(3) 喷墨打印机　　　　　　　　　(4) 多媒体计算机
(5) 客户机/服务器体系结构　　　　(6) 扩展卡

(7) 超链接 (8) 主存
(9) 辅存 (10) 颜色饱和度

Ⅱ. Translate the following key words into Chinese. (20%)

(1) Image projection technology (2) Cathode ray tube
(3) clock frequency (4) CD-ROM
(5) HTML (6) Distributed hypermedia
(7) Domain name system (8) TCP/IP
(9) video compression (10) WWW

Ⅲ. Multiple choice. (30%)

1. If we break the word multimedia into its component parts, we get multi—meaning more than one, and media—meaning form of communication. Those types of media include: Text, Audio Sound, Static Graphics Images, __(1)__, and Full-Motion Video. Text is the basis for __(2)__ programs and is still the fundamental information used in many multimedia programs. In fact, many multimedia applications are __(3)__ on the conversion of a book to a computerized form. This conversion gives the user immediate access to the text and lets him or her display pop-up windows, which give definitions of certain words. As a multimedia programmer, you can choose what font to display text in, how big(or small)it should be, and what color it should be displayed in. By __(4)__ text in more than one format, the message a multimedia application is trying to portray can be made more understandable. One type of application, which many people use every day, is the Windows Help Engine. This application is a text based information viewer that makes accessing information related __(5)__ a certain topic easy.

(1) A. Sound   B. Animation   C. pictures   D. Dynamic Images
(2) A. applications B. word processing C. communication D. drawing
(3) A. based   B. used   C. relied   D. depended
(4) A. observing B. searching   C. watching   D. displaying
(5) A. with   B. on   C. to   D. in

Answers: (1)_____ (2)_____ (3)_____ (4)_____ (5)_____

2. All parties involved in a communication must agree on a set of rules to be used when __(1)__ messages, including the language to be used and the rules for when messages can be sent. Diplomats call such an __(2)__ a protocol. The term is applied to computer communication as well-a set of rules that __(3)__ the format of messages and the appropriate actions required for each message is known as a network protocol or a computer communication protocol. The software that carries out such rules is __(4)__ protocol software. An individual network protocol can be as simple as an agreement to use ASCII when transferring a text file, or as complex as an agreement to use a complicated mathematical __(5)__ to encrypt data.

(1)~(5): A. function   B. exchanging   C. agreement   D. called   E. specify

Answers: (1)_____ (2)_____ (3)_____ (4)_____ (5)_____

3. The arithmetic/logic unit(ALU)is the functional unit that __(1)__ the computer with logical and computational capabilities. __(2)__ are brought into the ALU by the __(3)__, and the

ALU performs whatever arithmetic or logic operations are required to help carry out the instruction. (4) operations include adding, subtracting, multiplying, and dividing. Logic operations make a comparison and take action base on the results. For example, two numbers might be compared (5) determine if they are equal. If they are equal, processing will continue; if they are not equal, processing will stop.

  (1) A. provides   B. provided   C. provide   D. is providing
  (2) A. Codes   B. data   C. Bytes   D. Bits
  (3) A. logic unit   B. control unit   C. arithmetic/logic unit   D. arithmetic unit
  (4) A. Control   B. Logic   C. Arithmetic   D. Arithmetic/logic
  (5) A. with   B. of   C. to   D. for

Answers: (1)_____ (2)_____ (3)_____ (4)_____ (5)_____

Ⅳ. Translate the following four paragraphs into Chinese. (30%)

1. MP3, which stands for Motion Picture Expert Group (MPEG) —layer3, is rapidly becoming the standard for storing, trading, selling and stealing music via the Internet, according to analysts and audio experts. To understand how MP3 works, remember that sound travels in constantly changing waves. To save sound onto a PC or CD, a computer records snapshots (快照) of those waves at short intervals, a technique known as sampling. Playing samples reproduces the original sound; the more samples, the more realistic the sound.

2. UDP, User Datagram Protocol, is a pretty thin protocol in the sense that it does not add significant to the semantics of IP. It takes what IP provides and adds only two important features and one of them is optional. The first is port numbers which allow the sender to distinguish among multiple application programs on a given remote machine. The optional part of the UDP protocol is checksumming. This is a mechanism for determining whether part of the UDP data has been accidentally modified in transit.

3. ISDN stands for Integrated service Digital Network. The goal of ISDN is to replace the current telephone network, which requires digital-to-analog conversions, with facilities (设备,工具) totally devoted to (专用于) digital switching and transmission, yet advanced enough to replace traditionally analog forms of data, ranging from voice to computer transmissions, music, and video. ISDN is built on two main types of communications channels: a B channel, which carries data at a rate of 64 kbps, and a D channel, which carries control information at either 16 or 64 kbps. Computers and other devices connect to ISDN lines through simple, standardized interfaces. When fully implemented, ISND is expected to provide users with faster, more extensive services.

试卷二参考答案

Ⅰ．关键词翻译（汉译英）。(20%)

  (1) domain name server   (2) programming language   (3) ink-jet printer
  (4) multimedia computer   (5) client/server architecture   (6) expansion card

(7) hyperlink   (8) main memory   (9) secondary storage   (10) color saturation

Ⅱ．关键词翻译（英译汉）。**(20%)**

(1) 图像投影技术　(2) 阴极射线管　(3) 时钟频率　(4) 一次写多次读　(5) 超文本标识语言　(6) 分布式超媒体　(7) 域名系统　(8) 传输控制协议/因特网协议　(9) 视频压缩　(10) 万维网

Ⅲ．选择填空。**(30%)**

1．(1)B　(2)B　(3)A　(4)D　(5)C
2．(1)B　(2)C　(3)E　(4)D　(5)A
3．(1)A　(2)B　(3)B　(4)C　(5)C

Ⅳ．英译汉。**(30%)**

1. MP3，代表活动图像专家组第3层。据分析家和音频专家称，MP3迅速成为通过因特网存储、交换、销售和窃取音乐的标准。为了明白MP3如何工作，应记住声音是以不断变化的波形传播的。为了将声音录制在PC或CD上，计算机以很短的时间间隔记录那些声波的快照，这就是称为采样的技术。播放样本重新产生原声，采样越多，声音越逼真。

2. UDP即用户数据包协议，它十分简单，它没有对IP的内涵做重大扩展。它只是利用了IP协议的现成的规范，并增加了两个重要的功能，其中一项是可选的。第一项是端口号，发送方借此识别远程计算机上运行的多个应用程序。UDP协议的可选项是求和校验。这一机制用于判定UDP的部分数据在传输过程中是否被意外地修改。

3. ISDN即综合业务数字网，它是从现存的电话服务发展而来的全球性的数字通信网。它的目标是用完全专用于数字交换和传输的设备来代替当前需要进行数模转换的电话网，它已经足够先进，可以代替传统的数据模拟形式，这些数据包括声音、计算机传输、音乐和视频。ISDN基于两种主要类型的通信信道：以64kbps速率传输数据的B信道和以16或64kbps速率传输控制信息的D信道。计算机和其他设备通过简单标准的接口接入ISDN线路。当ISDN完全实现时，ISDN有望给用户提供更快、更广泛的服务。

试 卷 三

Ⅰ．**True or False. (5%)**

(1) ＿＿＿ Computer virus are small programs.

(2) ＿＿＿ The control unit is the functional unit that provides the computer with logical and computation capabilities.

(3) ＿＿＿ The only element that be deleted or removed is the one that was inserted most recently into a stack.

(4) ＿＿＿ A collection of interconnected computers is called internet.

(5) ＿＿＿ JPEG is an ISO/ITU standard for storing motion pictures in compressed digital form.

Ⅱ．**Translate the key words into Chinese. (20%)**

(1) Office Automation　　　　　　　　(2) Frequency Division Multiplexing

(3) proxy server
(4) Video On Demand
(5) CAD
(6) Artificial Intelligence
(7) DBMS
(8) cable television
(9) HTML
(10) ISDN

Ⅲ. **Translate the key words into English. (17%)**
(1) 专家系统
(2) 系统配置
(3) 分组交换
(4) 函数调用
(5) 视频压缩
(6) 协议栈
(7) 赋值语句
(8) 适配器
(9) 接口
(10) 算法

Ⅳ. **Fill in the blanks with appropriate computer terms. (20%)**

(1) _____ ,which stands as the heart, or, you may just as well say, the brain of the computer, carrying out instructions provided either by the user of the system itself.

(2) _____ ,which enables a computer to store, at least temporarily, data and program.

(3) _____ ,which covers such equipment as the display screen, the printer, and various other devices through which you can see what the computer has accomplished.

(4) _____ in a network, a device joining communication lines at a center location, providing a common connection to all devices on the network.

(5) _____ A device that converts data from one form into another, as from one form usable in data processing to another form usable in telephonic transmission.

(6) _____ An address for a resource on the Internet, which consists of the protocol, the name of the server and the path to a resource.

(7) _____ A computer-based text retrieval system that enables the user to provide access to or gain information related to a particular text.

(8) _____ the protocol used to copying files to and from remote computer systems on a network using TCP/IP such as the Internet.

(9) _____ , which, in the form of a keyboard and a mouse, is the channel through which data and instructions enter a computer.

(10) _____ A program that translates another program written in a high-level language into machine language so that it can be executed.

A. memory   B. output device   C. URL   D. CPU   E. Hub
F. input device G. modem   H. FTP   I. compiler   J. hypertext

Ⅴ. **Multiple choice. (20%)**

1. The management information system(MIS)concept has been defined in dozens of ways. Since one organization's model of an MIS is likely to __(1)__ from that of another, it's mot surprising that their MIS definitions would also vary in scope and breadth. For our purposes, and MIS can be defined as a __(2)__ of computer based data processing procedures developed in an organization and integrated as necessary with manual and other procedures __(3)__ the purpose of providing timely and effective information to support __(4)__ and other necessary management

functions. An organization is also divided vertically into different specialties and functions, which require separate information flows. Combing the horizontal managerial levels with the vertical specialties produces the complex organizational structure. Underlying this structure is a database consisting of internally and externally produced data relating to past, present, and predicted future events. The formidable task of the MIS designer is to develop the information flow needed to support decision making. Generally speaking, much of the information needed by managers who occupy different levels and who have different responsibilities is obtained from a collection of existing information systems(or subsystems). These systems may be tied together very closely in an MIS. More often, however, they are more (5) .

    (1) A. differ   B. different   C. keep   D. prevent

    (2) A. protocol   B. process   C. network   D. file

    (3) A. with   B. in   C. from   D. for

    (4) A. managerial level   B. information   C. decision making D. outcome

    (5) A. dependent   B. changeable   C. closely coupled D. loosely coupled

Answers: (1)_____ (2)_____ (3)_____ (4)_____ (5)_____

2. Multimedia aims at making computers (1) to use. As you can see some examples, the added information provided by sound and color picture will make getting to the destination faster. A multimedia application will usually store huge amounts of (2) (usually on CD-ROM), so that you can also get the same information for other communities. This type of information access makes computers much more (3) for the user. (4) adding multimedia to your programs, you can make computers more interesting and much more fun for the user. Interactive multimedia is another buzzword going through the industry. It is nothing (5) an application that gets input from the user. A book is mot very interactive, and neither is television. Because of the inclusion of a keyboard and a mouse, a multimedia computer is built for user input. That makes personal computers the premier interactive multimedia machines of today.

    (1) A. easier   B. more convenient   C. more difficult   D. broader

    (2) A. texts   B. information   C. data   D. files

    (3) A. interesting   B. suitable   C. valuable   D. different

    (4) A. For   B. with   C. In   D. By

    (5) A. of   B. rather   C. more than   D. but

Answers: (1)_____ (2)_____ (3)_____ (4)_____ (5)_____

3. All parties involved in a communication must agree on a set of rules to be used when (1) messages, including the language to be used and the rules for when messages can be sent. Diplomats called such an (2) a protocol. The term is applied to computer communication as well —— a set of rules that (3) the format of messages and the appropriate actions required for each message is known as a network protocol or a computer communication protocol. The software that carries out such rules is (4) protocol software. An individual network protocol can be as simple as an agreement to use ASCII when transferring a

text file, or as complex as an agreement to use a complicated mathematical (5) to encrypt data.

Words to be chosen from:

A. function   B. exchanging   C. agreement   D. called   E. specify

Answers: (1)_____ (2)_____ (3)_____ (4)_____ (5)_____

Ⅵ. Translate the following paragraph into Chinese. (10%)

Software is the sequences of instructions in one or more programming languages that comprise a computer application to automate some business function. Engineering is the use of tools and techniques in problem solving. Putting the two words together, software engineering is the systematic application of tools and techniques in the development of computer-based application.

Ⅶ. Reading comprehension. (8%)

There are four main types of viruses: shell, intrusive, operating system and source code.

Shell viruses wrap themselves around a host program and do not modify the original program. Shell programs are easy to write, which is why about half of all viruses are of this type. In addition, shell viruses are easy for programs like Data Physician to remove.

Intrusive viruses invade an existing program and actually insert a portion of themselves into the host program. Intrusive viruses are hard to write and very difficult to remove without damaging the host file.

Shell and intrusive viruses most commonly attack executable program files—those with a .com or .exe extension, although data files are also at some risk.

Operating system viruses work by replacing parts of the operating system with their own logic. Very difficult to write, these viruses have the ability, once booted up, to take total control of your system. According to Digital Dispatch, known versions of operating system viruses have hidden large amount of attack logic in falsely marked bad sectors. Others install Ram-resident programs or device drivers to perform infection or attack functions invisibly from memory.

Source code viruses are intrusive programs that are inserted into a source program such as those written in Pascal prior to the program being compiled. These are the least-common viruses because they are not only hard to write, but also have a limited number of hosts compared to the other types.

Questions:

1. Which viruses are the most common viruses? _____

 A. shell viruses   B. intrusive viruses

 C. operating system viruses   D. source code viruses

2. Which viruses wrap themselves around a host program and do not modify the original program? _____

 A. operating system viruses   B. source code viruses

 C. shell viruses   D. intrusive viruses

3. Intrusive viruses _____.

 A. are easy to write

 B. can be easily to remove without damaging the host file

C. most commonly attack executable program files
D. wrap themselves around a host program
4. Which statement is not true? _____
A. Shell viruses can be killed by programs like Data Physician.
B. Source code viruses are inserted into a source program after the program is being compiled.
C. operating system viruses work by replacing parts of the operating system with their own logic.
D. source code viruses are few than any other viruses.

试卷三参考答案

Ⅰ．判断。**(5%)**
(1) T   (2) F   (3) T   (4) F   (5) F

Ⅱ．关键词翻译（英译汉）。**(20%)**
(1) 办公自动化   (2) 频分多路复用   (3) 代理服务器   (4) 视频点播
(5) 计算机辅助设计   (6) 人工智能   (7) 数据库管理系统
(8) 有线电视   (9) 超文本标识语言   (10) 综合业务数据字网

Ⅲ．关键词翻译（汉译英）。**(17%)**
(1) expert system   (2) system configuration   (3) packet switching   (4) function call
(5) video compression   (6) protocol stack   (7) assignment statement   (8) adapter
(9) interface   (10) algorithm

Ⅳ．术语填空。**(20%)**
(1) D   (2) A   (3) B   (4) E   (5) G   (6) C   (7) J   (8) H   (9) F   (10) I

Ⅴ．选择填空。**(20%)**
1. (1) A   (2) C   (3) D   (4) C   (5) D
2. (1) A   (2) B   (3) C   (4) D   (5) D
3. (1) B   (2) C   (3) E   (4) D   (5) A

Ⅵ．英译汉。**(10%)**
软件是用一种或多种程序设计语言写的指令序列，它们构成了使某个商业功能自动化的计算机应用程序。工程是在问题解决过程中使用工具和技术。将这两个词合在一起，软件工程就是在开发基于计算机的应用程序中系统地应用工具和技术。

Ⅶ．阅读理解。**(8%)**
1. B   2. C   3. C   4. B

试 卷 四

Ⅰ．True or False. **(5%)**
(1) _____ Computer virus are living cells.

(2) _____ The ALU is the functional unit that provides the computer with logical and computation capabilities.

(3) _____ The only element that be deleted or removed is the one that was inserted most recently into a stack.

(4) _____ A collection of interconnected computers is called internet.

(5) _____ MPEG is an ISO/ITU standard for storing motion pictures in compressed digital form .

Ⅱ. Translate the key words into Chinese. (20%)

(1) color depth   (2) Time Division Multiplexing
(3) TCP/IP   (4) Integrated Circuit
(5) CAI   (6) neural network
(7) SQL   (8) DNS
(9) SMTP   (10) HDTV

Ⅲ. Translate the key words into English. (17%)

(1) 软件工程   (2) 虚拟现实
(3) 电路交换   (4) 递归函数
(5) 音频格式   (6) 信息技术
(7) 模式识别   (8) 路由器
(9) 索引   (10) 中断

Ⅳ. Fill in the blanks with appropriate computer terms. (20%)

(1) _____ ,which stands as the heart, or, you may just as well say, the brain of the computer, carrying out instructions provided either by the user of the system itself.

(2) _____ ,which enables a computer to store, at least temporarily, data and program.

(3) _____ ,which covers such equipment as the display screen, the printer, and various other devices through which you can see what the computer has accomplished.

(4) _____ in a network, a device joining communication lines at a center location, providing a common connection to all devices on the network.

(5) _____ A device that converts data from one form into another, as from one form usable in data processing to another form usable in telephonic transmission.

(6) _____ An address for a resource on the Internet, which consists of the protocol, the name of the server and the path to a resource.

(7) _____ A computer-based text retrieval system that enables the user to provide access to or gain information related to a particular text.

(8) _____ the protocol used to copying files to and from remote computer systems on a network using TCP/IP such as the Internet.

(9) _____ , which, in the form of a keyboard and a mouse, is the channel through which data and instructions enter a computer.

(10) _____ A program that translates another program written in a high-level language

into machine language so that it can be executed.

    A. memory   B. output device   C. URL   D. CPU   E. Hub

    F. modem   G. input device   H. FTP   I. compiler   J. hypertext

Ⅴ. Multiple choice. (20%)

1. A virus is a program or piece of code that is __(1)__ onto your computer without your knowledge and runs against your wishes. Most virus can also replicate themselves. All computer viruses are __(2)__. A simple virus that can make a copy of itself over and over again is relatively easy to produce. Even such a simple virus is __(3)__ because it will quickly use all available memory and bring the system to a halt. An even more dangerous type of virus is one capable of transmitting itself across networks and bypassing (绕过、避开) security systems.

Since 1987, when a virus infected APARNET, a large network used by the Defense Department and many universities, many anti-virus programs have become __(4)__. These programs periodically __(5)__ your computer system for the best-known types of viruses.

    A. available   B. check   C. dangerous   D. loaded   E. man-made

Answers: (1)_____ (2)_____ (3)_____ (4)_____ (5)_____

2. Multimedia aims at making computers to __(1)__ use. As you can see some examples, the added information provided by sound and color picture will make getting to the destination faster. A multimedia application will usually store huge amounts of __(2)__ (usually on CD-ROM), so that you can also get the same information for other communities. This type of information access makes computers much more __(3)__ for the user. __(4)__ adding multimedia to your programs, you can make computers more interesting and much more fun for the user. Interactive multimedia is another buzzword going through the industry. It is nothing __(5)__ an application that gets input from the user. A book is mot very interactive, and neither is television. Because of the inclusion of a keyboard and a mouse, a multimedia computer is built for user input. That makes personal computers the premier interactive multimedia machines of today.

(1) A. easier   B. more convenient   C. more difficult   D. broader

(2) A. texts   B. information   C. data   D. files

(3) A. interesting   B. suitable   C. valuable   D. different

(4) A. For   B. with   C. In   D. By

(5) A. of   B. rather   C. more than   D. but

Answers: (1)_____ (2)_____ (3)_____ (4)_____ (5)_____

3. Technically, the WWW is a __(1)__ hypertext system that uses the Internet __(2)__ its transport mechanism. In a hypertext system, you __(3)__ by clicking hyperlinks, which display another document which also contains __(4)__. What makes the Web such an exciting and useful medium is that the next document you see could be housed on computer next door or half-way around the world. The Web makes the Internet easy to use. Created in 1989 at a research institute in Switzerland, the Web relies upon the hypertext transport protocol(HTTP), an Internet standard that __(5)__ how an application can locate and acquire resources stored on another computer on

the Internet.

    A. specifies   B. global   C. hypertext   D. as   E. navigate

Answers: (1)_____ (2)_____ (3)_____ (4)_____ (5)_____

Ⅵ. Translate the following paragraphs into Chinese. (10%)

Data that is stored more-or-less permanently in a computer we term a database. The software that allows one or many persons to use and/or modify this data is a database management system (DBMS). The primary goal of a DBMS is to provide an environment that is both convenient and efficient to use in retrieving information from and storing information into the database.

Ⅶ. Reading Comprehension. (8%)

Input to the computer and output from the computer are provided by peripheral devices which have many different purposes. Most essential of all is the console terminal which provides communication between the user and the computer. Typically, this is a keyboard plus either a video display device or a printing type output device. Other peripheral devices include communication equipments which permit transfer of data over telephone lines to and from distant computers.

In a lab or in a manufacturing environment analog signals from voltage and current sources as well as electrical signals indicating pressure or temperature can be digitized by an Analog to Digital converter and, presented to the computer as binary numbers. Similarly binary data from the computer can be converted to an analog voltage by a Digital to Analog converter. Digital recording of music uses these principles.

Graphics terminals permit interactive presentation of information in the form of bar graphs and maps as well as drawings for electronic and mechanical parts. Mechanical parts can be designed using a graphics terminal, then mathematically subjected to stress analysis to determine weak points. A modified design can then be produced in hours rather than days or weeks. All this is possible through the use of these specialized peripheral devices plus the supporting software or programs to control them. Computer Assisted Design encompasses the above processes and equipment.

Questions:

1. The typical peripheral devices do not include _____.

    A. keyboard   B. printer   C. video display   D. telephone

2. Analog signals _____.

    A. can be converted to binary number

    B. can be digitized by a Digital to Analog converter

    C. can be digitized by a video converter

    D. can be digitized by a keyboard

3. Which statement is true? _____

    A. Analog signals can not indicate temperature.

    B. Graphics terminals prohibit interactive presentation of information in the form of bar graphs and maps.

C. The output of Digital to Analog converter is analog signal.

D. Computer Assisted Design do not need any peripheral device.

4. The title of this paragraph may be _____.

A. input /output device B. Analog signal

C. Peripheral Devices D. Computer Communication

试卷四参考答案

Ⅰ．判断。**(5%)**

(1) F　　(2) T　　(3) T　　(4) F　　(5) T

Ⅱ．关键词翻译（英译汉）。**(20%)**

(1) 颜色深度　(2) 时分多路复用　(3) 传输控制协议/网际协议　(4) 集成电路
(5) 计算机辅助教学　(6) 神经网络　(7) 结构化查询语言　(8) 域名系统
(9) 简单邮件传输协议　(10) 高清电视

Ⅲ．关键词翻译（汉译英）。**(17%)**

(1) software engineering　(2) virtual reality　(3) circuit switching　(4) recursive function
(5) audio format　(6) information technology　(7) pattern recognition　(8) router
(9) index　(10) interrupt

Ⅳ．术语填空。**(20%)**

(1) D　(2) A　(3) B　(4) E　(5) F　(6) C　(7) J　(8) H　(9) G　(10) I

Ⅴ．选择填空。**(20%)**

1. (1) D　(2) E　(3) C　(4) A　(5) B（1分1个）
2. (1) A　(2) B　(3) C　(4) D　(5) D（1分1个）
3. (1) B　(2) D　(3) E　(4) C　(5) A（1分1个）

Ⅵ．英译汉。**(10%)**

我们把不同程度地长久存储在计算机中的数据称为数据库。允许一个或多个人使用或修改该数据的软件称为数据库管理系统。DBMS 的主要目的是提供这样的环境：从数据库中检索信息和把信息存储在数据库中时使用起来既方便又高效。

Ⅶ．阅读理解。**(8%)**

1. D　　2. A　　3. C　　4. B

附录三　软件水平考试程序员级专业英语试题节选

1. 从供选择的答案中选出应填入下列英语文句中____内的正确答案，把编号写在答卷的对应栏内。

Software products may be __A__ into four basic types: application programs, programming language processors, operating systems, and system utilities.

Application programs are programs that __B__ useful tasks such as solving statistical problems, or keeping your company's books.

Programming language processors are programs that __C__ the use if a computer language in a computer system. They are tools for the development of application programs.

Operation systems are programs that __D__ the system resources and enable you to run application programs.

System utilities are special programs that __E__ the usefulness of or add capabilities to a computer.

供选择的答案

A～E：　(1) manage　　(2) perform　　(3) support　　　(4) reduce
　　　　(5) divided　　(6) enhance　　(7) implemented　(8) introduce
　　　　(9) ranked　　 (10) run

答案：A. (5)　　B. (2)　　C. (3)　　D. (1)　　E. (6)

2. 从供选择的答案中选出应填入下列英语文句中____内的正确答案，把编号写在答卷的对应栏内。

Here is a useful procedure for choosing a program:

1. Study the features of all the programs you might choose __A__. Decide which features you need, which you would __B__, and which you can do your jobs without.

2. Eliminate the programs that clearly do not __C__ you needs.

3. Consider how the remaining programs perform the functions you will use most often. This can affect a program's usability more than all the "nice" features that you will __D__ need.

4. Study the remaining programs carefully—with __E__ experience if you can get it—and decide which one is best for you.

供选择的答案

A：(1) for　　(2) on　　(3) in　　(4) from
B，C：(1) meet　(2) require　(3) help　(4) give　(5) choose　(6) like
D，E：(1) often　(2) seldom　(3) always　(4) rich　(5) hands-on　(6) little

答案：A. (4)　　B. (6)　　C. (1)　　D. (2)　　E. (5)

3. The C programming language has __A__ one of the most __B__ programming languages, and it has been implemented on most personal computers and multiuser systems, especially those designed for research and development. It evolved from the version described in Kernighan and Ritche's work (called "K&R C" after the authors) into __C__ variants, including the standard ANSI C, which __D__ many type-checking features and includes a standard library. Of the two main __E__, K&R C is probably the most commonly used on multiuser computers, with ANSI C close behind; in the personal computing world, ANSI C is far more common.

供选择的答案

A～E: (1) much (2) variants (3) complex (4) incorporates
 (5) several (6) become (7) popular (8) editions
 (9) come (10) users

答案：A. (6)   B. (7)   C. (5)   D. (4)   E. (2)

4. The use of the computer is changing the very __A__ of many jobs that exist within a business. In the industrial __B__, tools were developed to assist in improving production, but much work still involved __C__ labor. The information __B__ has brought about another change—a change from __C__ labor to __D__ labor.

Pressure on computer worker can be great. Whether operating a robot, running a computer, or programming a computer, a single error can be __E__. The smallest error could misdirect an airline, disrupt delivery schedules, or cost millions of dollars.

供选择的答案

A～E: (1) important (2) great (3) disastrous (4) physical
 (5) body (6) brain (7) mental (8) revolving
 (9) revolution (10) nature

答案：A. (10)   B. (9)   C. (4)   D. (7)   E. (3)

5. 从供选择的答案中选出应填入____内的正确答案，把编号写在答卷的对应栏内。

For years, users toiling under the 640 KB __A__ memory constraints of MS-DOS have suffered severe memory __B__ problems. Help is available now from DOS extenders. This software technique enables MS-DOS programs to access up to 16 MB of __C__ memory on an 80286-based PC and up to 4 GB on an 80386-based PC. __D__ release 3.0 from Microsoft Corp. is the most widely publicized package to use a DOS __E__.

供选择的答案

A～E: (1) extender (2) expanded (3) internal (4) argument
 (5) conventional (6) management (7) protected (8) X-window
 (9) Windows (10) security

答案：A. (5)   B. (6)   C. (3)   D. (9)   E. (1)

6. 从供选择的答案中选出应填入____内的正确答案，把编号写在答卷的对应栏内。

The UNIX system contains several __A__ that comply with the definition of a software tool. Among them are programs that __B__ and manipulate text, programs that analyze text files, and programs that format text files to produce high quality hard copy suitable for __C__.

One characteristic of these tools is that they operate on ordinary test __D__, which means that you can read the input and output files by simply listing them on a __E__.

供选择的答案

A~E: (1) terminal   (2) keyboard   (3) programs   (4) programming
      (5) files   (6) directories (7) create   (8) build
      (9) publication (10) painting

答案: A. (3)   B. (7)   C. (9)   D. (5)   E. (1)

7. Applications put computers to practical business __A__, but below the __B__ it's the heart of an operating system—the kernel—that provides the technical wizardry to juggle multiple programs, connect to networks and store __C__.

A traditional kernel provides all the functions for applications. The kernel __D__ memory, I/O devices and parcels out processor time.

The kernel also supports security and fault __E__, which is the ability to recover automatically when parts of the system fail.

供选择的答案:

A: (1) used   (2) use   (3) apply   (4) applied
B: (1) earth   (2) bottom   (3) table   (4) surface
C: (1) graphics (2) data   (3) text   (4) image
D: (1) manages   (2) manage   (3) managed   (4) managing
E: (1) error   (2) question (3) tolerance   (4) problem

答案: A. (2)   B. (4)   C. (2)   D. (1)   E. (3)

8. By using MP3, a 600M-byte music CD can be __A__ to 50M bytes or less. It can be streamed (downloaded in chunks) so that you can begin listening to the opening bars while the __B__ of the file arrives in the backgrounD. And, most important, MP3 music files retain good listening __C__ that __D__ compression schemes lacked. That __E__ of features makes accessing and distributing music on the Web practical for the first time.

供选择的答案:

A: (1) pressed   (2) compressed (3) compress   (4) press
B: (1) past   (2) next   (3) rest   (4) host
C: (1) amount   (2) mass   (3) quantity   (4) quality
D: (1) earlier   (2) front   (3) later   (4) backward
E: (1) addition (2) combination (3) difference (4) condition

答案: A. (2)   B. (3)   C. (4)   D. (1)   E. (2)

9. software design is a __A__ process. It requires a certain __B__ of flair on the part of the designer. Design can not be learned from a book. It must be practiced and learnt by experience and study of existing systems. A well __C__ software system is straightforward to implement and maintain, easily __D__ and reliable. Badly __C__ software systems, although they may work are __E__ to be expensive to maintain, difficult to test and unreliable.

A: (1) create	(2) created	(3) creating	(4) creative
B: (1) amount	(2) amounted	(3) mount	(4) mounted
C: (1) design	(2) designed	(3) designing	(4) designs
D: (1) understand	(2) understands	(3) understanding	(4) understood
E: (1) like	(2) likely	(3) unlike	(4) unlikely
答案：A. (4)	B. (1)	C. (2)	D. (4)	E. (2)

10. The error messages given by a C compiler show the message text, the most common cause of the error, and a suggestion for____the error.

    A. updating	B. fixing	C. changing	D. editing

11. The following suggestions increase "programs" (1) and make them easier to (2) :

① Use a standard indention technique, blank lines, form feeds, and spaces.

② Insert plenty of comments into your code.

(1) A. reliability	B. security	C. readability	D. usability

(2) A. execute	B. interrupt	C. compile	D. Maintain

12. A____is a feature of the system or a description of something the system is capable of doing in order to fulfill the system's purpose.

    A. plan	B. requirement	C. document	D. design

13. 100BASE-TX makes use of two pairs of twisted pair cable, one pair used for transmission and the other for____.

    A. reception	B. detection	C. relation	D. connection

14. Each instruction is processed sequentially, and several instructions are at varying stages of execution in the processor at any given time, this is called instruction____.

    A. executing	B. sequencing	C. pipelining	D. producing

15. One use of networks is to let several computers share____such as file systems, printers, and tape drives.

    A. CPU	B. memory	C. resource	D. data

16. A floating constant consists of an integer part, a decimal point, a fraction part, an e or E, and an optionally signed integer____.

    A. exponent	B. order	C. superfluous	D. superior

17. The (1) program means a program written in a high-level language. It is generally translated to an (2) program, which is in a form directly understandably by the computer. The translation is usually done by a program called (3) .

(1) A. assemble	B. web	C. C	D. source

(2) A. object	B. basic	C. C	D. assemble

(3) A. compiler	B. assembler	C. compile	D. transfer

18. ____processing offers many ways to edit text and establish document formats. you can easily insert, delete, change, move and copy words or blocks of text.

    A. Data	B. Database	C. Word	D. File

19. "scrolling" is a technique most commonly associated with____.

    A. disk   B. display   C. printe   D. memory

  20. Firewall is a____mechanism used by organizations to protect their LANs from the Internet.

    A. reliable   B. stable   C. peaceful   D. security

  21. A____consists of the symbols, characters, and usage rules that permit people to communicate with computer.

    A. programming language B. network   C. keyboard   D. display

  22. In C language, a____is a series of characters enclosed in double quotes.

    A. matrix   B. string   C. program   D. stream

  23. In C language,_____ are used to create variables and are grouped at the top of a program block.

    A. declarations   B. dimensions   C. comments   D. descriptions

  24. An____statement can perform a calculation and store the result in a variable so that it can be used later.

    A. executable   B. input   C. output   D. assignment

  25. Each program module is compiled separately and the resulting____files are linked together to make an executable application.

    A. assembler   B. source   C. library   D. object

  26. A floating constant consist of an integer part, a decimal point, a fraction part, and e or E, and an optionally signed integer _____.

    A. exponent   B. order   C. superfluous   D. superior

  27. ____is the address of a variable or a variable in which the address of another variable is stored.

    A. Directory   B. Pointer   C. Array   D. Record

  28. ____is non-program text embedded in a program to explain its form and function to human readers.

    A. Command   B. Compile   C. Comment   D. Statement

  29. ____is a sequence of letters and digits, the first character must be a letter.

    A. An identifier   B. A string   C. An array   D. Program

  30. The error messages given by a C compiler shoe the message text, the most common cause of the error, and a suggestion fo____the error.

    A. updating   B. fixing   C. changing   D. editing

  31. In C language, one method of communicating data between functions is by____.

    A. arguments   B. variables   C. messages   D. constants

  32. In C language, all variables must be (1) before use, usually at the beginning of the function before any (2) statements.

    (1) A. stated   B. instructed   C. illustrated   D. declared

    (2) A. operative   B. active   C. executable   D. processing

  33. When a string constant is written in C program, the compiler creates____of characters

containing the characters of the string, and terminating it with "\0".

    A. a group   B. an array   C. a set   D. a series

34. In C language, __(1)__ variables have to be defined outside function, this __(2)__ actual storage for it.

    (1) A. internal   B. output   C. export   D. external

    (2) A. locates   B. allocates   C. finds   D. looks for

35. In C language, the increment and decrement____can only be applied to variables, so an expression like x=(i+j)++ is illegal.

    A. operation   B. operate   C. operator   D. operand

36. In C program, it is convenient to use a____to exit form a loop.

    A. end   B. break   C. stop   D. quit

37. In C language, ____is a collection of one or more variables, possibly of different types, grouped together under a single name for convenient handling.

    A. a structure   B. a file   C. and array   D. a string

38. In C language, the usual expression statements are____or function calls.

    A. I/Os   B. assignments   C. operations   D. evaluations

39. ____is a set of specifications and software that allow small programs or software components to work together.

    A. ActiveX   B. XML   C. HTML   D. DBMS

40. An____statement can perform a calculation and store the result in a variable so that it can be used later.

    A. executable   B. input   C. output   D. assignment

41. A sequence of any number of characters enclosed in the double quotes " " is called a character____.

    A. array   B. group   C. set   D. string

答案：

10. B 11. (1) C (2) D 12. B 13. A 14. C 15. C 16. A 17.(1) D (2) A (3) A
18. C 19. B 20. D 21. A 22. B 23. A 24. D 25. D 26. A 27. B 28. C 29. A
30. B 31. A 32. (1) D (2) C 33. B 34. (1) D (2) B 35. A 36. B 37. A 38. D
39. A 40. D 41. D

附录四　英语构词法和基本句型

一、构词法（Word Formation）

1. 转化法

动词—名词	rest	have a rest
名词—动词	mail	mail a postcard
形容词—动词	slow	slow down the car
形容词—名词	final	the tennis final

2. 合成法

1）合成形容词

形容词+名词+ed	good tempered, noble-minded
形容词+现在分词	good looking, fine-sounding
副词+现在分词	hard-working, far-reaching
名词+现在分词	peace-loving, epoch-making
副词+过去分词	well-known, widespread
形容词+名词	large-scale, high-level
名词+形容词	duty-free, airsick
其他	first-rate, all-round

2）合成名词

名词+名词	database, keyboard
形容词+名词	hardware, software
动名词+名词	waiting-room, swimming-pool
动词+名词	payroll, typewriter
名词+动名词	handwriting, sunbathing
动词+副词	get-together, breakthrough
副词+动词	output, input
其他	well-being, good-for-nothing

3）合成动词

名词+动词	sleepwalk, eavesdrop
介词+动词	overflow, undergo
形容词+动词	whitewash, blacklist

4）其他合成词（副词、代词等）

maybe, myself, nevertheless

3. 派生法

即在原词基础上加上适当的前缀或后缀构成一个新词的方法。

prefix + word　　word + suffix

1）前缀

前缀	含义/作用	词 例	词 义
ab-	偏离，脱出，离开	abnormal, abuse	反常的，滥用
anti-	反，抗；防止；非，对立；非传统，非正统	anticompetitive, anticancer, anticorrosion, antiart	反竞争，抗癌，防腐蚀，非传统的艺术
auto-	自己，本身；自己的，自动的	autobiography, autograph, automobile, automation, autonomy	自传，亲笔签名，汽车，自动化，自治
e-	出去	emigrate, emit, eminent, elect	移居国外，射出，杰出的，选举
bio-	生命；生物；生物学的	biography, biology, biochemistry	传记，生物，生物化学
ex-	构成名词 以前的	ex-wife, ex-president, export	前妻，前总统，出口
ex-	向外，向上	export, exit, exclude	输出，出口，不包括
en/em-	构成动词，可用于形容词或名词之后	enlarge, enrich, endanger, encourage, empower	扩大，使丰富，使处于……危险之中，鼓励，授权给
extra-	一般加在形容词前，表示在……之外，越出，超出	extraordinary, extraspecial, extracurricular	不平常的，临时增刊的，课外的
be-	视作；使，使显得	befriend, belittle, beware	以朋友相待，轻视，当心
bi-	二，双（倍）	biannual, bicycle	两年一次或一年两次，自行车
co-	共同；联合；同事，伙伴	coexist, cooperate, collaborate, coauthor, copilot	共存，合作，协作，合著者，副驾驶
con-	一起，和……	confuse, confer, conclude, concord	使混乱，授予，得出结论，和谐
counter/contra-	反，逆；相反的；反方向的；对应，对等	counteract, counterattack, counterplot, contradiction, counterpart	抵制，反攻，对抗策略，矛盾，对应的人或物
dis-	不；除去，分离，剥夺	disagree, disbelieve, disregard, disarm, disperse, disinterest	不同意，不相信，忽视，解除武装，使消散，公正无私
de-	除去；离开，脱离；贬，降低	debone, derail, devalue	去掉骨头，使脱轨，使货币贬值
fore-	在前；早先，预先；前部	foreleg, forefather, foreshadow, foresee	前腿，祖先，预示，预见
hyper-	超出，在……之上，过度，（存在于）三维空间以上	hyperactive, hypersensitive, hyperplane	活动过强的，过敏的，超平面
in/im-	在……里，在……上；[构成反义词]不	include, immigrate, invisible, instable, inappropriate	包括，从国外移入，看不见的，不稳定的，不合适的
inter-	互相；在……之间，在……内	interact, interchange, intercourse, intercultural	相互作用，交换，往来，不同文化间的
il-/ir-	构成反义词，不	illegal, illogic, iliterate, irregular, irrelevant	不合法的，不符合逻辑的，文盲，没有规律的，无关的
im-	构成反义词，不	impossible, immeasurable	不可能的，不可估量的
macro-	大的，长的，宏观的，大规模的	microclimate, microeconomics, microstructure	大气候，宏观经济学，宏观结构

英语构词法和基本句型

续表

前缀	含义/作用	词例	词义
mal-	坏，恶，不良；不正当，非法	malpractice, malnutrition, maladminister, maladjustment	渎职，营养不良，对……管理不善，调节不良
micro-	微小的，极小的	microwave, microsoft, microcomputer	微波，微软，微型计算机
mis-	坏，不当；错；不，缺少，相反	misdeed, mismanage, misspell, misjudge, mistrust	不端行为，对……管理不善，错误的拼写，错误的判断，不信任
mini-	极小的，微型的，极短的	miniature, minibike, minibus, minicab	微小的模型，微型摩托车，中客车，微型出租车
mono-/uni	单一的	monotone, monoplane, uniform	单调的语调，单翼机，一致的
multi/poly	多的，多倍的；多方面的	multifunction, multination, multicultural, multiplayer	多功能的，多国的，多种文化的，多层的
non-	非，无	non-English, non-agricultural non-dairy, non-contact	非英语，非农业的，不含奶的，没有接触的
out-	在外，向外；出；结果；超过，胜过	outhouse, outpatient, outgrow	外屋，门诊病人，过度成长
over-	外面的；上面的；越过；太，过度	overcoat; overhang; overweight, overwork, overestimate	外衣，悬挂于……上，超重，过度工作，过高估计
post-	在……之后	postwar, postgraduate, postdoctoral	战后时期，研究生，博士后研究人员
pre-	先，预先，前面	prereview, prearrange, preadolescence	预习，预先安排，青春前期的少年
pro-	亲，赞成	pro-American, protagonist	亲美的，拥护者
pseudo-	伪，假	pseudoscience, pseudonym, pseudoplastic	伪科学，假名，假塑料的
psycho-	精神，灵魂，心理	pshchoanalysis, psychobiology, psychodynamic	精神分析，精神生物学，心理历史学，精神动力的
re-	再次，重	return, rebuild, reconsider	归来，重建，重新考虑
semi/hemi-	半；部分地，不完全地	semifinal, semiprofessional, semieducated, hemisphere	半决赛（的），半职业性的，受过部分教育的，半球
self-	自身的；为自身的；	selfassuarance, selfesteem	自信，自信心
sub-	下面；进一步；次要，从属；次于，亚于；	submarine, subsoil, subdivide,	潜艇，下层土，再分
super-	上，上方；进一步，次要；过，过分；超，超级	superimpose, superstructure	叠加，上层建筑
trans-	横穿，横断；在……的那边；超越，胜过	transcontinental, transplant	横跨大陆的，移植
tri-	三，三次；每三……一次	triangle, trisect, trialogue, trimonthly	三角（形），把……分成三等份，三人（方）谈，每三个月一次的
un-	做……相反的动作；使自由，松开；不	undress, undo, unconscious	脱去衣服，撤销，未意识到的
under-	在……下面，低于，次于	underground, undercharge, underclothe, undergraduate	在地面下，对……要价（或收费）过低，给……穿内衣，大学肄业生

2）后缀

后缀	含义/作用	词 例	词 义
-an	表示人	American, African	美国人，非洲人
-ant	表示人或物	servant, assistant, participant, consultant, accountant, pollutant, immigrant, attendant	仆人，助手，参加者，顾问，会计，污染物，外来移民，侍从
-ar	表示人	scholar, liar, beggar	学者，说谎者，乞丐
-ee	表示人	employee, trainee, payee, examinee	雇员，受训者，受款者，受试者
-eer	表示人	mountaineer, volunteer engineer	爬山运动员，志愿者，工程师
-ent	表示人	correspondent, respondent, student	通信者，调查对象，学生
-er	表示人	employer	雇主
-ese	表示一个国家的人或语言	Chinese, Japanese, Portugese	中国人（汉语），日本人（日语），葡萄牙人（葡萄牙语）
-ese	表示女性，动物雌性	actress, waitress, lioness	女演员，女侍者，母狮
-or	表示人	actor, sailor, governor, conductor, educator, survivor, translator	演员，海员，州长，售票员，教育家，幸存者，翻译者（家）
-acy	表示性质、状态或情况	tendency, infancy, sufficiency, efficiency, accuracy	倾向，婴儿期，足够，效率，精确（性）
-age	构成名词，表示结果、行动或状态	shortage, storage, marriage, postage	缺乏，储存，婚姻，邮资
-al	构成名词，表示行动或动作	refusal, arrival, approval, denial, dismissal, criminal	拒绝，到达，同意，否认，解散，罪犯
-al	构成形容词	moral, emotional, conventional, envi-ronmental, incidental, principal, educational	道德的，情感的，常规的，环境的，偶然的，主要的，教育的
-ance/ ancy-	表示动作、状态或特性	guidance, acquaintance, disturbance, appearance	指导，认识，打扰，出现
-able/ ible	能够……的，适于……，值得……的；易受……的	divisible, inevitable, reasonable	可分的，不可避免的，合理的
-ate	构成动词	enumerate, accumulate, formulate, generate, aggravate, stimulate, exaggerate, decorate	列举，积累，阐明，产生，加重，鼓励，夸张，装饰
-ce	构成名词	difference, dependence, persistence, patience, convenience	不同，依靠，坚持，耐心，方便
-dom-	领域，领土，版图	freedom, kingdom, officialdom, bordom	自由，王国，官场，厌烦
-en	用于形容词或名词之后，构成动词意思是使变成	deepen, widen, shorten, soften, darken, redden, brighten, hearten, heighten	加深，拓宽，缩短，使变软，变黑，变红，变亮，鼓舞，加高
-ence/ -ency	表示性质或状态	confidence, existence, frequency emergency	信心，存在，频率，紧急情况

续表

后缀	含义/作用	词例	词义
-ery	地点或行为	nursery, robbery	幼儿园，抢劫
-ful	充满……的；有……特色的；	eventful, careful, hopeful, painful, colorful	多事的，仔细的，有希望的，疼痛的，丰富多彩的
-hood	表示状态或时期	childhood, manhood, neighborhood	童年时期，成人时期，邻居
-ian	表示人	musician, technician, physician, historian, electrician	音乐家，技术员，内科医生，历史学家，电工
-ion/tion	表示情况、状态、性质、行为	fashion, introduction, consumption, realization, modernization	时髦，介绍，消费，实现，现代化
-ic/ical	关于……的，属于……的	realistic, optimistic, toxic, pessimistic, historical political	现实的，乐观的，有毒的,悲观的，历史上的，政治上的
-ics	……学	statistics, electronics, economics, physics	统计学，电子学，经济学，物理学
-ify	构成动词	simplify, clarify, modify, identify, classify	简化，澄清，修饰，识别，分类
-ly	如……的；有……特性的，可构成形容词或副词	motherly, brotherly, lovely, manly	母亲般的，兄弟般的，可爱的，男子汉气的
-ish	属于……的，(贬义)有……特性的，有些……的	foolish, childish, selfish	愚笨的，孩子般的，自私的
-ism	……主义	socialism, capitalism	社会主义，资本主义
-ist	表示人，可指授权做谋事的人，(思想、主义、学说的)信奉者	typist, scientist, agriculturist, journalist	打字员，科学家，农学家，记者
-ity	构成名词	superiority, similarity, majority, minority, priority, personality, nationality	优越(性)，相似点，多数，少数，优先(权)，个性，国籍
-ive	构成形容词	effective, competitive, cooperative, exclusive, intensive, negative	有效的，竞争的，合作的，排除的，集中的，否定的
-ize	构成动词	specialize, emphasize, memorize, analyze, sympathize, maximize, minimize	专门研究，强调，记住，分析，同情,扩大到最大限度，降低在最小限度
-less	没有……的	meaningless, jobless, endless, hopeless	没有意义的，失业的，没有终止的，没有希望的
-man	表示某一行为的实施者	fishman, milkman, spaceman, fireman, businessman	钓鱼的人，送牛奶的人，宇航员，消防队员，商人
-ment	构成名词，表示行为的原因、方法；结果	argument, advertisement, requirement, assessment, instrument	争辩，广告，要求，评估，仪器
-mum	构成名词	maximum, minimum, optimum	最大限度，最小限度，最佳

续表

后缀	含义/作用	词 例	词 义
-ness	表示状况、性质或程度	deafness, neatness, carelessness, illness, tenderness, shyness	聋，整洁，粗心，疾病，柔软，羞怯
-ous/eous/ious	构成形容词	dangerous, advantageous, industrious, nutritious	危险的，有利的，勤奋的，有营养的
-ship	表示状态或性质；职位身份或地位	hardship, friendship, leadership, membership, relationship	艰难，友谊，领导，会员，关系
-some	易于……的	troublesome, tiresome, burdensome, handsome, worrisome	麻烦的，疲劳的，累赘的，清秀的
-t	构成名词	complaint, threat, pursuit, trait	抱怨，威胁，追求，特性
-th	构成名词，表示状态、情形、过程等	growth, death, strength, length, width, depth, height	生长，死亡，力量，长度，宽度，深度，高度
-ure	构成名词	departure, pressure	离开，压力
-ty	构成名词	difficulty, certainty, loyalty, specialty, safety	困难，肯定，忠诚，专业，安全

4. 缩略法

顾名思义，缩略法就是将烦冗复杂的一些专业名词或组织的名称的首字母提取出来组成一个简短的新词。如：

UN——United Nations（联合国）

Ph. D.——Doctor of Philosophy

UFO——Unknown Flying Object（不明飞行物）

此外，一些网络流行语以及日常对话中也常用缩略法将对话精简，其中有部分缩略词已为大多数人所接受。与上述专业名词不同，这些词常用于非正式场合。如：DIY——do it yourself，BBS——Bulletin Board System。

二、数量的翻译

1. 整数的翻译

数幂	前缀	缩写	英文	译名	例子
10^1	deca-	Da	ten	十	decade
10^2	hecto-	H	hundred	百	hectogram
10^3	kilo-	K	thousand	千	kilometer(km), kilogram(kg)
10^6	mega-	M	million	百万，兆	megabyte(MB), megahertz(MHz)
10^9	giga-	G	billion	十亿，千兆，吉	gigabyte(GB), gigahertz(GHz)
10^{12}	tera-	T	trillion	万亿，兆兆，太	terabyte(TB)

2. 小数的翻译

数幂	前缀	缩写	英文	译名	例子
10^{-1}	deci-	d	tenth	十分之一，分	decimeter(dm)
10^{-2}	centi-	c	hundredth	百分之一，厘	centimeter(cm)

英语构词法和基本句型

续表

数幂	前缀	缩写	英文	译名	例子
10^{-3}	milli-	m	thousandth	千分之一，毫	millimeter(mm), milligram(mg)
10^{-6}	micro-	u	millionth	微	microwave, microcomputer
10^{-9}	nano-	n	billionth	毫微，纳	nanometer（nm），nanosecond(ns)
10^{-12}	pico-	p	trillionth	微微，皮可	picosecond(ps), picofarad(pf)

三、非谓语动词的用法

1. 不定式

1）作主语

To ignore this would be a mistake.

To see is to believe.

2）作宾语

I like to make new friends.

You can learn how to type there.

3）作表语

The problem is to find a solution.

What worries me is to speak at the meeting.

4）作定语

It's time to get up.

There's nothing to worry about.

5）作宾语补足语

I expect you to be at the station on time.

Try to persuade him to come with us.

6）作状语

He is silly to do such a thing.

You are kind to look after the baby for me.

I'm surprised to hear that.

2. 动名词

1）作主语

Painting is his hobby.

2）作表语

His hobby is painting.

3）作宾语

I can't help thinking so.

I'm looking forward to returning home.

4）作定语

This is the air cooling passage.

There is a living room.

3. 分词

现在分词表示"主动，进行"，过去分词表示"被动，完成"。

1）作定语

单个分词作定语时，相当于一个形容词，直接置于被修饰词的前面，翻译时采用顺译法。

PASCAL is a well-designed language.

分词短语作后置定语，翻译时应置于名词之前，若定语很长，则可分出来译成一个分句。

In effect, the screen often called a monitor, serve as a window on main memory, allowing the user to view its contents.

There is a piano standing in the corner.

2）作表语

The theory sounds quite convincing.

This software is not well-suited for commercial use.

3）作宾语补足语

I'm sorry to have kept you waiting.

He couldn't make himself believed.

You'd better have the computer repaired.

4）作状语

She turned up, dressed in pink.

I'm very busy recently, having no time to play tennis.

Nobody having anymore to say, the meeting was closed.

Seen from the space, the Great Wall is like a dragon.

There being nothing to do, she went home.

四、英语的五种基本句型结构

句子由主语和谓语两大部分组成。主语结构比较单一，谓语结构则不然，不同类别的谓语动词导致不同的谓语结构，从而形成了不同的句型（Sentence Pattern）。换句话说，不同的句型是由不同类别的谓语动词所决定的，因此，句型又被称为动词句型（Verb Pattern）。语法家们对句型的分类不尽相同，一般认为，现代英语的基本句型主要有五种。

1. 句型一：Subject（主语）+Verb（谓语）

这种句型中的动词大多是不及物动词，所谓不及物动词，就是这种动词后不可以直接接宾语。常见的动词如：work, sing, swim, fish, jump, arrive, come, die, disappear, cry, happen 等。如：

（1）Li Ming works very hard.李明学习很努力。

（2）The accident happened yesterday afternoon.事故是昨天下午发生的。

2. 句型二：Subject（主语）+Link. V（系动词）+Predicate（表语）

这种句型主要用来表示主语的特点、身份等。其系动词一般可分为下列两类：

（1）表示状态。这样的词有：be, look, seem, smell, taste, sound, keep 等。如：

① This kind of food tastes delicious.这种食物吃起来很可口。

② He looked worried just now. 刚才他看上去有些焦急。

（2）表示变化。这类系动词有：become, turn, get, grow, go 等。如：

① Spring comes. It is getting warmer and warmer. 春天到了，天气变得越来越暖和。

② The tree has grown much taller than before. 这棵树比以前长得高多了。

3. 句型三：Subject（主语）+Verb（谓语）+Object（宾语）

这种句型中的动词一般为及物动词，所谓及物动词，就是这种动词后可以直接接宾语，其宾语通常由名词、代词、动词不定式、动名词或从句等来充当。例：

（1）He took his bag and left.（名词） 他拿着书包离开了。

（2）Li Lei always helps me when I have difficulties.（代词）当我遇到困难时，李雷总能给我帮助。

（3）She plans to travel in the coming May Day.（不定式）她打算在即将到来的"五一"外出旅游。

（4）I don't know what I should do next.（从句）我不知道下一步该干什么。

注意：英语中的许多动词既是及物动词，又是不及物动词。

4. 句型四：Subject（主语）+Verb（谓语）+Indirect object（间接宾语）+Direct object（直接宾语）

这种句型中，直接宾语为主要宾语，表示动作是对谁做的或为谁做的，在句中不可或缺，常常由表示"物"的名词来充当；间接宾语也被称之为第二宾语，去掉之后，对整个句子的影响不大，多由指"人"的名词或代词承担。引导这类双宾语的常见动词有：buy, pass, lend, give, tell, teach, show, bring, send 等。如：

（1）Her father bought her a dictionary as a birthday present. 她爸爸给她买了一本词典作为生日礼物。

（2）The old man always tells the children stories about the heroes in the Long March. 老人经常给孩子们讲述长征途中那些英雄的故事。上述句子还可以表达为：

（1） Her father bought a dictionary for her as a birthday present.

（2） The old man always tells stories about the heroes to the children in the Long March.

5. 句型五：Subject（主语）+Verb（动词）+Object（宾语）+Complement（补语）

这种句型中的"宾语+补语"统称为"复合宾语"。宾语补足语的主要作用或者是补充、说明宾语的特点、身份等，或者表示让宾语去完成的动作等。担任补语的常常是名词、形容词、副词、介词短语、分词、动词不定式等。如：

（1）You should keep the room clean and tidy. 你应该让屋子保持干净整洁。（形容词）

（2）We made him our monitor.（名词）我们选他当班长。

（3）His father told him not to play in the street.（不定式）他父亲告诉他不要在街上玩。

（4）My father likes to watch the boys playing basketball.（现在分词）

（5）Yesterday I had a picture taken with two Americans.（过去分词）

- 常见的动词有：tell, ask, advise, help, want, would like, order, force, allow 等。
- 注意：动词 have, make, let, see, hear, notice, feel, watch 等后面所接的动词不定式作宾语补足语时，不带 to。如：

（1）The boss made him do the work all day. 老板让他整天做那项工作。

（2）I heard her sing in the next room all the time last night. 昨天晚上我听见她在隔壁唱了一个晚上。

五、独立主格结构

1. 独立主格结构的构成

1）名词/代词+分词（现在分词、过去分词）+主句

Time permitting, we can finish the work.

The signal given, the bus started.

The question being settled, we all feel excited.

2）主句+名词/代词+形容词/副词

The children were making a snowman, hands red with cold.

The boy looked at the picture, eyes wide open.

3）名词/代词+形容词/副词，+主句

The meeting over, the students were dismissed.

Everying (being ready), they started out.

4）主句+名词/代词+不定式

He invited us to see a film, he himself to buy the tickets.

The mid-term exam is over, the end-of-term exam to come two months later.

5）名词/代词+不定式+主句

The teacher to help us, we will succeed.

For more students to see the example, the teacher showed it around the classroom.

6）主句，名词/代词+介词短语

He entered the dark room, gun in hand.

He came in, book in hand.

The old peasant came back, a large basket on his shoulder.

2. 独立主格结构的特点

（1）独立主格结构的逻辑主语与句子的主语不同，它独立存在。

（2）名词或代词与后面的分词、形容词、副词、不定式、介词等是主谓关系。

（3）独立主格结构一般有逗号与主句分开。

六、科技英语的特点与翻译

科技英语目前已经发展成为一种重要的语体，属于英语中的一大类，亦称非文学类，它泛指一切论及或谈及科学和技术的书面语和口语，大致有专题著作、专题论文、教科书、实验报告和方案、各类科技情报和文字资料、科技使用手册的结构描述和操作规程、新闻报告等等。

1. 科技英语的句法特点与翻译

首先，广泛使用长句。科技英语与科学技术和自然现象有关，更强调事实的准确性和逻辑性，因此，要求严谨精确，而它使用的长句常是一个主句后接几个从句，从句也可带短语，构成极其复杂，这就增加了阅读和翻译的难度。

其次，非谓语动词使用较多。使用非谓语动词可简化句子结构，使句子更紧凑连贯，

它的三种形式 to、V-ing 和 V-ed，可以帮助译者确定各个成分之间的逻辑关系。在科技英语中，有代表性的用法是分词词组做状语，这能直观地看出它们之间的逻辑关系，翻译时可以给译者提供很大帮助。最后，科技翻译强调的是事实或现象，而不是人，所以，经常使用被动语态。例如，不知道或不必说明行为者时，可以用被动语态；对行为对象比对行为者更感兴趣时，也可使用被动语态。在翻译时，我们可以灵活地将被动句译为主动句或判断句。虽然被动语态的句子比较呆板和单一，但科技英语中却不得不使用。

其特点如下：

（1）语态上，广泛使用被动语态和无人称句。

科学技术强调的是事物、行动过程、客观变化和自然规律，不允许个人的思想感情等主观因素来歪曲客观真相，而被动语态能客观地描述事实。

例 1. We use computer memory to store date tempararily.

为了强调客观事实，使用被动语态将使表达更得体：

例 2. Computer memory is used to temporarily store data.

科技英语中强调的是动作、状态和客观事实，而不是动作的执行者，所以在表达上一般用：

例 3. It took Professor Wang 5 months to develop a new machine.

来替代：

Professor Wang spent 5 months in developing a new machine.

（2）句子结构上，长句使用较多。

在表达科学技术的一些复杂概念时，只有借助结构较复杂的句子才能清楚地表达各种主从关系、逻辑关系以及意义上的不同层次。

例 4.In reality,computer memory is only capable of remembering sequences of zeros and ones,but by utilizing the binary number system it is possible to produce arbitray rational numbers and through clever formatting all manner of representations of pictures,sounds,and animations.

这个句子看似复杂，其实是个简单句，主语是 produce arbitray rational numbers 和 representations of pictures,sounds,and animations,谓语是 is possible。it 是形式主语。

（3）行文风格上，科技语言力求平易和精确翻译，尽力避免使用旨在加强语言感染力和宣传效果的各种修辞，忌用夸张、借喻、讥讽、反诘、双关及押韵等修辞手法，以免使读者产生行文浮华、内容虚饰之感。

（4）词汇含义上，一些词组的含义与生活用语中的含义相去甚远。

在计算机英语中，bus 为总线，bridge 为网桥，cache 为高速缓存，level 为电平，character 为字符，process 为进程，thread 为线程，switch 为交换机，它们在公共英语中分别为公共汽车、桥、地窖、水平、性格、处理或过程、线、开关。

总之，翻译前要认真阅读文章，深入分析句子结构，能够真正理解逻辑关系，翻译时采取有效的翻译策略，而不是翻译成中式英语或欧式汉语。

2. 科技英语的文本特点与翻译

科技英语翻译与其他文本翻译的不同之处，在于其本身具有整体的专业性和极强的规范性。它不像文学翻译，可以适当地再创作，结合文化因素，帮助读者更好地理解译文。也不像广告翻译，具有较强的趣味性和吸引力。做科技翻译时，可以从其他文本翻译理论中得到启发。掌握词、句子的特点和翻译方法后，整体把握篇章就容易了。翻译一些特殊文本时，也要注意，如翻译产品说明书，要简洁明了，方便理解，如果译得特别专业，反而晦涩难懂，阅读时会给译文的读者带来很大难度。

　　科技英语的翻译方法不是唯一的，不同的文本需要采用不同的翻译方法，甚至有时不可能只使用其中的某一种方法。应针对科技英语的不同特点，采取对应有效的翻译策略。

附录五　学术英语写作常用句子

Before you start:

(1) Pay close attention to the words **in bold**, which are often used in conjunction with the main word.

(2) [ ] means "insert a suitable word here", while () means "this word is optional".

(3) Bear in mind that, within each group, some examples are slightly more formal / less frequent than others.

- **Debate**

1. [X] has **fostered debate on** _____. (fostered=encouraged)

2. There has been **an inconclusive debate about whether** _____.

3. The question of whether _____ has **caused much debate** in [our profession][over the years].

4. (Much of) **the current debate revolves around** _____.

- **Discussion**

1. In this section/chapter, the **discussion** will **point to** _____.

2. The **foregoing discussion** implies that _____. (foregoing=that came before)

3. **For the sake of discussion**, I would like to argue that _____.

4. In this study, **the question under discussion** is _____.

5. in this paper, **the discussion centers on** _____.

6. [X] **lies at the heart of the discussion on** _____.

- **Evidence** (Remember: Evidence is uncountable)

1. The **available evidence** seems to suggest that _____ / point to _____.

2. **On the basis of** the **evidence** currently available, it seems fair to suggest that _____.

3. There is **overwhelming evidence** corroborating the notion that _____. (corroborating=confirming)

4. **Further evidence** supporting / against [X] may lie in the findings of [Y], who _____.

5. These results **provide confirmatory evidence** that _____.

- **Ground**

1. I will now summarize the **ground covered** in this [chapter] by _____.

2. **On logical grounds**, there is no compelling reason to argue that _____.

3. [X] **takes a middle-ground position** on [Y] and argues that _____.

4. **On these grounds**, we can argue that _____.

5. [X]'s views are **grounded on the assumption** that _____.

- **Issue**

1. This study is an attempt to **address the issue of** _____.

2. In the present study, **the issue under scrutiny** is _____.

3. The **issue of whether** _____ is **clouded** by the fact that _____. (clouded=made less clear)

4. To **portray the issue** in [X]'s terms, _____.

5. Given the **centrality of this issue to** [my claim], I will now _____.

6. This [chapter] is **concerned with the issue of** [how/whether/what] _____.

- **Literature**

1. [X] is **prominent in the literature on** [Y].

2. There is a **rapidly growing literature on** [X], which indicates that _____.

3. The **literature shows no consensus on** [X], which means that _____.

4. The (current) **literature on [X] abounds with examples of** _____.

- **Premise**

1. The main **theoretical premise behind** [X] is that _____.

2. [X] and [Y] **share** an important **premise**: _____.

3. [X] is **premised on the assumption** that _____.

4. The **basic premises of** [X]'s theory / argument are _____.

5. The **arguments against** [X]'s **premise** rest on [four] assumptions.

- **Research**

1. This study **draws on research** conducted by _____.

2. Although there has been relatively little **research on / into** [X], _____.

3. In the last [X] years, [educational] **research** has **provided ample support for** the assertion that _____.

4. **Current research** appears / seems to **validate the view** that _____.

5. **Research on / into** _____ does not **support the view** that _____.

6. **Further research in this area** may include _____ and _____.

7. Evidence for [X] is **borne out by research** that shows _____.

8. There is **insufficient research on /into** _____ to draw any firm conclusion about / on _____.

- **View**

1. The **consensus view** seems to be that _____.

2. [X] **propounds the view** that _____. (propound=put forward for consideration)

3. Current research (does not) appear(s) to **validate** such a **view**.

4. There has been **dissenters to the view** that _____. (dissenter=someone who disagrees)

5. The answer to [X] / The difference between [X] and [Y] is not as clear-cut as **popular views** might suggest.

6. The **view** that _____ is (very much) **in line with** [common sense].

7. I am **not alone in my view** that _____.

8. [X] **puts forward the view** that _____.

9. [X]'s **views rest on the assumption** that _____.

附录六　词汇表

单　词　表

abbreviate	[ə'bri:vieit]	vt. 缩略；使简短；缩简
abbreviation	[ə.bri:vi'eiʃn]	n. 省略，缩写，简化，缩写词
address	[ə'dres]	n. 地址；称呼；演说；通信处
agency	['eidʒənsi]	n. 代理；机构
algorithm	['ælgəriðəm]	n. 算法；运算法则
alignment	[ə'lainmənt]	n. 队列；排成直线；对齐
alphanumeric	[,ælfənju:'merik]	adj. 文字数字的
analog	['ænəlɔ:g]	n. 类似物；模拟　adj. 模拟的
analyze	['ænəlaiz]	vt. 分析；分解
animation	[,æni'meiʃn]	n. 动画片
applet	['æplət]	n. Java 小应用程序
arbitrary	['ɑ:bitrəri]	adj. 随意的，任性的
array	[ə'rei]	n. 数组；队列；阵列
assign	[ə'sain]	vt. 分派，选派，分配；归于
associate	[ə'səuʃieit]	vt. 联想；(使)发生联系；(使)联合；
associative	[ə'səuʃiətiv]	adj. 联合的，联想的
atom	['ætəm]	n. 原子；原子能
attachment	[ə'tætʃmənt]	n. 附件，附属物
attribute	[ə'tribju:t]	vt. 认为……是；把……归于　n. 属性；特征
audio	['ɔ:diəu]	adj. 音频的；听觉的
author	['ɔ:θə(r)]	n. 作者　vt. 创作出版
authority	[ɔ:'θɔrəti]	n. 权威；权力
authorization	[,ɔ:θərai'zeiʃn]	n. 授权，批准
background	['bækgraund]	n. 背景；底色；背景资料；配乐
bandwidth	['bændwidθ]	n. 带宽
beep	[bi:p]	n. 哔哔声　v. 嘟嘟响
behavior	[bi'heivjə]	n. 行为；态度
beneath	[bi'ni:θ]	prep. 在……的下方

biometrics	[ˌbaiəuˈmetriks]	n. 生物识别技术
blank	[blæŋk]	n. 空白表格，填空处　adj. 空白的
board	[bɔːd]	n. 板；董事会 vt. 上（船、车或飞机）
boldface	[ˈbəuldfeis]	n. 黑体字，粗体
browser	[ˈbrauzə(r)]	n. 浏览器
bundle	[ˈbʌndl]	n. 捆；一批
bus	[bʌs]	n. [计]总线；公共汽车
cable	[ˈkeibl]	n. 绳索；电缆
calculation	[ˌkælkjəˈleʃən]	n. 计算；盘算；估计
capability	[ˌkeipəˈbiləti]	n. 性能；容量；才能；能力
cart	[kɑːt]	n. 运货马车，手推车
cartoon	[kɑːˈtuːn]	n. 漫画；动画片
centimeter	[ˈsentiˌmiːtə]	n. 厘米
characteristic	[ˌkærəktəˈristik]	n. 特性，特征，特色
chat	[tʃæt]	vi. 聊天；闲谈
chip	[tʃip]	n. 芯片
circuit	[ˈsɜːkit]	n. 巡回；电路；线路
client	[ˈklaiənt]	n. 顾客；客户机
collision	[kəˈliʒn]	n. 碰撞；冲突
color	[ˈkʌlə]	n. 颜色，色彩；
command	[kəˈmɑːnd]	n. 命令，指挥；指令
compile	[kəmˈpail]	vt. 编译；编制；汇编
compiler	[kəmˈpailə(r)]	n. 汇编者；编译程序
component	[kəmˈpəunənt]	n. 成分；零件；[数]要素　adj. 组成的
compression	[kəmˈpreʃn]	n. 压缩；压紧
conference	[ˈkɔnfərəns]	n. 会议；讨论
connector	[kəˈnektə(r)]	n. 连接器，连接体
container	[kənˈteinə(r)]	n. 容器；箱
contextual	[kənˈtekstʃuəl]	adj. 上下文的，前后关系的
control	[kənˈtrəul]	vt. 控制；管理；
convention	[kənˈvenʃn]	n. 规矩；惯例，习俗
conversation	[ˌkɔnvəˈseiʃn]	n. 交谈，会话
converter	[kənˈvɜːtə(r)]	n. 变换器；变压器
cursor	[ˈkɜːsə(r)]	n. 光标
customize	[ˈkʌstəmaiz]	vt. 定制，定做
debugger	[ˌdiːˈbʌgə(r)]	n. 调试器
decode	[ˌdiːˈkəud]	vt. 译（码），解（码）

decryption	[diːˈkripʃn]	n. 解密，译码
dedicate	[ˈdedikeitid]	adj. 专用的；专注的
default	[diˈfɔːlt]	n. 默认值，缺省　adj. 默认的
definition	[ˌdefiˈniʃn]	n. 定义；规定 [物]清晰度
delete	[diˈliːt]	vt. & vi. 删除
demodulator	[diːˈmɔdjuleitə]	n. 解调器；检波器
depth	[depθ]	n. 深度；深处
desktop	[ˈdesktɔp]	n. 桌面
destination	[ˌdestiˈneiʃn]	n. 目的，目的地，终点
detection	[diˈtekʃn]	n. 侦查；检查
determine	[diˈtɜːmin]	vt. 决定，确定
dialog box	[ˈdaiəlɔɡ bɔks]	n. 对话框
digital	[ˈdidʒitl]	adj. 数字的
dimension	[diˈmenʃən, dai-]	n. 尺寸；[数]次元，度，维
discipline	[ˈdisəplin]	n. 纪律；学科；训练
discrete	[diˈskriːt]	adj. 分离的，离散的
disk drive	[disk draiv]	n. 磁盘驱动器
distribute	[diˈstribjuːt]	vt. 分配；散布；分发
document	[ˈdɔkjumənt]	n. 文档，证件
domain	[dəˈmein]	n. 范围，领域；域
dominate	[ˈdɔmineit]	v. 支配；影响
don	[dɔn]	vt. 穿上，披上
download	[ˌdaunˈləud]	v. 下载
dynamic	[daiˈnæmik]	adj. 动态的；动力的，动力学的
earphone	[ˈiəfəun]	n. 耳机，听筒
efficiency	[iˈfiʃnsi]	n. 功效；效率
electrical	[iˈlektrikl]	adj. 用电的，与电有关的，电学的
emulation	[ˌemjuˈleiʃn]	n. 竞赛；仿效
enclose	[inˈkləuz]	vt. 把……围起来；把……装入信封
encryption	[inˈkripʃn]	n. 加密；编密码
endpoint	[ˈendpɔit]	n. 端点，终点
environment	[inˈvairənmənt]	n. 环境，外界
equivalent	[iˈkwivələnt]	adj. 相等的；等价的　n. 对等物
erase	[iˈreiz]	vt. 清除；擦掉
execute	[ˈeksikjuːt]	vt. 执行；完成
expand	[ikˈspænd]	vt. 扩张；使……变大　vi. 扩展
expansion	[ikˈspænʃən]	n. 扩张；扩大；扩展

explicitly	[ik'splisitli]	adv. 明白地，明确地
expression	[ik'spreʃn]	n. 表现，表示，表达式
extension	[ik'stenʃn]	n. 伸展，扩大
feed	[fi:d]	vt. 喂养；向……提供
fiber	['faibə]	n. 光纤；纤维
field	[fi:ld]	n. 字段；场地，领域
filter	['filtə(r)]	n. 滤波器；滤光器；滤镜；过滤器　vt. 过滤
flash	[flæʃ]	n. 闪光　vt. & vi. 使闪光，使闪烁
flat	[flæt]	n. 平面；公寓　adj. 平的；单调的
folder	['fəuldə(r)]	n. 文件夹
font	[fɔnt]	n. 字体；字形
foreground	['fɔ:graund]	n. 前景
format	['fɔ:mæt]	n. 版式；形式
forward	['fɔ:wəd]	vt. 转寄；发送
frame	[freim]	n. 框架；边框
frequency	['fri:kwənsi]	n. 频率，次数
function	['fʌŋkʃn]	n. 功能，作用；函数
gateway	['geitwei]	n. 入口；网关
general-purpose	['dʒenrəl'pɜ:pəs]	adj. 多方面的；通用的
geographic	[ˌdʒi:ə'græfik]	adj. 地理学的，地理的
glove	[glʌv]	n. 手套
goggle	['gɔgl]	n. 护目镜；凝视
gray-scale	[grei skeil]	n. 灰度
handwriting	['hændraitiŋ]	n. 书法；笔迹
hierarchical	[ˌhaiə'rɑ:kikl]	adj. 分层的；等级（制度）的
home page	[həum peidʒ]	n. 主页，首页
host	[həust]	n. 主机；主人，东道主
hub	[hʌb]	n. 轮轴；中心；集线器
hyperlink	['haipəliŋk]	n. 超链接
hypermedia	[ˌhaipə'mi:diə]	n. 超媒体
icon	['aikɔn]	n. 图标；图符
identification	[aiˌdentifi'keiʃn]	n. 认同；鉴定，识别；验明
identifier	[ai'dentifaiə(r)]	n. 标识符
identify	[ai'dentifai]	vt. 确定；识别；认出　vi. 确定；认同
illegal	[i'li:gl]	adj. 非法的，违法的
illusion	[i'lu:ʒn]	n. 错觉；幻想；假象
image	['imidʒ]	n. 图像；肖像

impact	[ˈimpækt]	n. 影响；冲击 vt. 撞击；压紧
implement	[ˈimpliment]	vt. 实施，执行 n. 工具，器械
inch	[intʃ]	n. 英寸
incompatibility	[ˌinkəmˌpætəˈbiləti]	n. 不兼容；不一致
indent	[inˈdent]	n. （印刷中的）缩进；订单
index	[ˈindeks]	n. 索引；[数]指数；指示
individual	[ˌindiˈvidʒuəl]	adj. 个人的；个别的
individually	[ˌindiˈvidʒuəli]	adv. 分别地；各个地
infrastructure	[ˈinfrəstrʌktʃə(r)]	n. 基础设施；基础建设
ingredient	[inˈgri:diənt]	n. 因素；组成部分；要素
inherit	[inˈherit]	vt. & vi. 继承
input	[ˈinput]	n. & vt. 输入
insertion	[inˈsɜ:ʃn]	n. 插入（物）
install	[inˈstɔ:l]	vt. 安装；安置
institution	[ˌinstiˈtju:ʃn]	n. 机构；制度
instruction	[inˈstrʌkʃən]	n. 授课；指令
integrate	[ˈintigreit]	vt. 使一体化；使整合 vi. 成为一体
intensity	[inˈtensəti]	n. 强度；烈度
interface	[ˈintəfeis]	n. 界面；[计]接口
interfere	[ˌintəˈfiə(r)]	vi. 干预；干涉；打扰
interpretation	[inˌtɜ:priˈteiʃn]	n. 理解；解释，说明
interpreter	[inˈtɜ:pritə(r)]	n. 解释者；口译译员；解释程序
intranet	[ˈintrənet]	n. 内联网
invert	[inˈvɜ:t]	vt. 使……前后倒置；使反转
jet	[dʒet]	n. 喷气式飞机；喷嘴 vt. 喷射，喷出
justify	[ˈdʒʌstifai]	vt. 证明……有理；vi. 整理版面
keyboard	[ˈki:bɔ:d]	n. 键盘；琴键
kit	[kit]	n. 成套用品；配套元件
laser	[ˈleizə]	n. 激光
layout	[ˈleiaut]	n. 布局；安排；设计
legal	[ˈli:gl]	adj. 法律的；合法的
library	[ˈlaibrəri]	n. 库；图书馆
linear	[ˈliniə(r)]	adj. 直线的；[数]一次的；线性的
loop	[lu:p]	n. 回路；圈；循环
mainframe	[ˈmeinfreim]	n. 大型机；主机
manipulate	[məˈnipjuleit]	vt. 操纵；操作，处理
margin	[ˈmɑ:dʒin]	n. 边缘；页边距

marquee	[mɑːˈkiː]	n. 选取框；大帐篷；跑马灯
material	[məˈtiəriəl]	n. 素材；材料，原料
matrix	[ˈmeitriks]	n. [数]矩阵；模型
measurement	[ˈmeʒəmənt]	n. 量度；尺寸
mechanical	[miˈkænikəl]	adj. 机械的，机械学的
memory	[ˈmeməri]	n. 记忆；[计]存储器；内存
mesh	[meʃ]	n. 网眼，网格
metal	[ˈmetl]	n. 金属；金属元素
metaphor	[ˈmetəfə(r)]	n. 象征；隐喻；暗喻
mice	[mais]	n. 老鼠（的名词复数）；鼠标
microprocessor	[ˌmaikrouˈprousesə]	n. 微处理器
minicomputer	[ˈminikəmpjuːtə(r)]	n. 小型计算机
mining	[ˈmainiŋ]	n. 采矿；挖掘
mobile	[ˈməubail]	adj. 可移动的　n. 手机
modem	[ˈməudem]	n. 调制解调器
modulator	[ˈmɔdjuleitə]	n. 调制器；调节器
molecule	[ˈmɔlikjuːl]	n. 分子
monitor	[ˈmɔnitə(r)]	n. 显示屏　vt. 监督；监控　vi. 监视
monochrome	[ˈmɔnəkrəum]	n. 单色画，黑白照片　adj. 单色的，黑白的
motherboard	[ˈmʌðəbɔːd]	n. 主板，母板
motion	[ˈməuʃn]	n. 运动；手势；动机
mouse	[maus]	n. 鼠标；老鼠
mouseover	[maus ˈəuvə(r)]	n. 悬停；鼠标经过
multimedia	[ˌmʌltiˈmiːdiə]	n. 多媒体
nanotechnology	[ˌnænəutekˈnɔlədʒi]	n. 纳米技术，毫微技术；
natural	[ˈnætʃrəl]	adj. 自然的；天生的
neural	[ˈnjuərəl]	adj. 神经的
node	[nəud]	n. 结点
nonprofit	[ˌnɔnˈprɔfit]	adj. 非营利的
object	[ˈɔbdʒikt]	n. 物体；目标；对象
operand	[ˈɔpərænd]	n. 操作数；运算数
operation	[ˌɔpəˈreiʃn]	n. 手术；操作；[数]运算
operator	[ˈɔpəreitə(r)]	n. 操作员；运算符
optic	[ˈɔptik]	adj. 光学的；眼睛的
optimal	[ˈɔptiməl]	adj. 最佳的，最优的
option	[ˈɔpʃn]	n. 选项；选择权
output	[ˈautput]	n. & vt. 输出

override	[ˌəuvəˈraid]	vt. 覆盖；推翻
package	[ˈpækidʒ]	vt. 包装 n. 包裹；包
panel	[ˈpænl]	n. 面板；控制板
parallel	[ˈpærəlel]	adj. 平行的；并行的
parameter	[pəˈræmitə(r)]	n. 参数；决定因素
pared-down	[ˈpeədd'aun]	adj. 压缩的；简化的
participant	[pɑːˈtisipənt]	n. 参加者，参与者
pattern	[ˈpætn]	n. 模式；图案；花样
peripheral	[pəˈrifərəl]	adj. 外围的
phrase	[freiz]	n. 成语；说法；短语
pixel	[ˈpiksl]	n. 像素
plastic	[ˈplæstik]	n. 整形；塑料制品
plotter	[ˈplɔtə(r)]	n. 绘图仪
plug	[plʌg]	n. 插头；塞子 vt. & vi. 插入；塞住
pointer	[ˈpɔintə]	n. 线索；指针
pointing	[ˈpɔintiŋ]	n. 指示；磨尖
port	[pɔːt]	n. 港口；接口；端口
portray	[pɔːˈtrei]	n. 描绘；描述；画像
prediction	[priˈdikʃn]	n. 预言，预测
presentation	[ˌpreznˈteiʃn]	n. 陈述，报告
private	[ˈpraivət]	adj. 私有的，秘密的
procedure	[prəˈsiːdʒə(r)]	n. 程序，手续；过程，步骤
profile	[ˈprəufail]	n. 侧面，轮廓；人物简介
protocol	[ˈproutəkɔːl]	n. 礼仪；（数据传递的）协议
punctuation	[ˌpʌŋktʃuˈeʃən]	n. 标点符号
queue	[kjuː]	n. 队列
recognition	[ˌrekəgˈniʃn]	n. 认识，识别
record	[ˈrekɔːd]	n. 唱片；记录
reference	[ˈrefrəns]	n. 参考（书）；提及 v. 引用；参照
regenerate	[riˈdʒenəreit]	vt. & vi. 回收；使再生
register	[ˈrɛdʒistə]	n. 记录；登记簿；登记 vt. & vi. 登记；注册
removal	[riˈmuːvl]	n. 除去；搬迁
remove	[riˈmuːv]	vt. 去除；开除
repeater	[riˈpiːtə(r)]	n. 中继器
replicate	[ˈreplikeit]	vt. 复制，复写
represent	[ˌrepriˈzent]	vt. 表现，象征；代表，代理
ring	[riŋ]	n. 戒指，指环

router	[ˈruːtə(r)]	n.	路由器
routine	[ruːˈtiːn]	n.	常规；例行程序
routing	[ˈruːtɪŋ]	vt.	按某路线发送，路由
saturation	[ˌsætʃəˈreɪʃn]	n.	饱和度
screen	[skriːn]	n.	屏幕；银幕
scroll	[skrəul]	vt.	滚动；卷页
security	[sɪˈkjuərətɪ]	n.	安全；保证
segment	[ˈsegmənt]	n.	环节；部分；网段
semiconductor	[ˌsemɪkənˈdʌktə(r)]	n.	半导体
sensitivity	[ˌsensəˈtɪvətɪ]	n.	敏感；灵敏性
sensory	[ˈsensərɪ]	adj.	感觉的，感官的
separate	[ˈsepəreɪt]	vt. & vi.	分开；（使）分离；隔开
sequence	[ˈsiːkwəns]	n.	顺序；[数]数列，序列
shorthand	[ʃɔːt hænd]	n.	速记
signify	[ˈsɪgnɪfaɪ]	vt.	意味；预示
silicon	[ˈsɪlɪkən]	n.	[化]硅；硅元素
simulation	[ˌsɪmjuˈleɪʃn]	n.	模仿，模拟
simultaneous	[ˌsɪmlˈteɪnɪəs]	adj.	同时的
simultaneously	[ˌsɪməlˈteɪnɪəslɪ]	adv.	同时地；一齐
socket	[ˈsɔkɪt]	n.	插座；灯座；窝，穴
source	[sɔːs]	n.	根源，本源
speaker	[ˈspiːkə(r)]	n.	扬声器
specification	[ˌspesɪfɪˈkeɪʃn]	n.	规格，说明书；详述
spot	[spɔt]	n.	地点，场所
stack	[stæk]	n.	堆栈，垛；栈
static	[ˈstætɪk]	adj.	静止的，静态的
strip	[strɪp]	n.	长条；条板；带状地带
style sheet	[staɪl ʃiːt]	n.	样式表
stylus	[ˈstaɪləs]	n.	唱针，尖笔
subroutine	[ˈsʌbruːtiːn]	n.	子程序
subscript	[ˈsʌbskrɪpt]	n.	下标，脚注
substitute	[ˈsʌbstɪtjuːt]	n.	用……替代 vi. 替代物；代替者
supercomputer	[ˈsuːpəkəmpjuːtə(r)]	n.	超级计算机，巨型计算机
switch	[swɪtʃ]	n.	开关；转换；交换机
symbol	[ˈsɪmbl]	n.	符号；象征；标志
synonym	[ˈsɪnənɪm]	n.	同义词
synonymous	[sɪˈnɔnɪməs]	adj.	同义词的；同义的

synthesizer	[ˌsinθiˈsaizə]	n. 合成物；合成器
tablet	[ˈtæblət]	n. 药片；平板电脑
tag	[tæg]	n. 标签
telecommunication	[ˌtelikəˈmjuːniˈkeiʃn]	n. 电信
template	[ˈtempleit]	n. 模板；样板
thermal	[ˈθəːməl]	adj. 热的，温热的
thread	[θred]	n. 线；线程
topology	[təˈpɔlədʒi]	n. 拓扑结构；拓扑（学）
trace	[treis]	vt. 跟踪，追踪；追溯
track	[træk]	vt. 跟踪；追踪
traffic cop	[ˈtræfik kɔp]	n. 交通警察
transaction	[trænˈzækʃn]	n. 交易，业务，事务
transform	[trænsˈfɔːm]	vt. 变换；改变 vi. 改变
transmission	[trænsˈmiʃn]	n. 传送；播送
twisted	[ˈtwistid]	v. 扭，搓，缠绕
typeface	[ˈtaipfeis]	n. 字体；字样
unauthorized	[ʌnˈɔːθəraizd]	adj. 未授权的；未经许可的
unique	[juˈniːk]	adj. 仅有的；独一无二的
utility	[juːˈtiləti]	n. 实用程序；功用；效用；
validation	[ˌvæliˈdeiʃn]	n. 确认
variable	[ˈveəriəbl]	n. 可变因素，变量 adj. 变化的
variant	[ˈveəriənt]	n.（词等的）变体；变量
variation	[ˌveəriˈeiʃn]	n. 变化，变动
vector	[ˈvektə(r)]	n. 矢量，向量
version	[ˈvɜːʃn]	n. 版本；译文
video	[ˈvidiəu]	n. 录像；录像磁带；录像机
virtual	[ˈvɜːtʃuəl]	adj. 实质上的；虚拟的
virtualization	[vɜːtʃuəlaiˈzeiʃn]	n. 虚拟化
virus	[ˈvairəs]	n. 病毒；计算机病毒
volume	[ˈvɔljuːm]	n. 体积；卷；音量
wavelength	[ˈweivleŋθ]	n. 波段；波长
web page	[web peidʒ]	n. 网页
website	[ˈwebsait]	n. [通信]网站
word wrap	[wəːd ræp]	n. 自动换行

短 语 表

3D surround sound	三维环绕音效
abbreviated from	是……的缩写
analog signal	模拟信号
animation and video	动画与视频
application sharing	应用共享
applications software	应用软件
array element	数组元素
associative array	关联数组
atom and molecule	原子与分子
audio and video	音频与视频
be referred to as	被称为
binary tree	二叉树
biometric data	生物数据
blank space	空白
built in	内置的
bus network	总线网
cable line	电缆线
Client/Server Computing	客户机/服务器计算
cloud computing	云计算
cloud storage	云存储
code editor	代码编辑程序
collision detection	冲突检测
color depth	色深度
color saturation	色彩饱和度
commercial business	商业企业
communicate with	与……通信
compression formation	压缩结构
computer monitor	计算机显示器
computer science	计算机科学
computer virus	计算机病毒
contextual menu	上下文菜单，快捷菜单
continuous motion	连续运动
control buses	控制总线
control panel	控制面板

customized tag	定制标签
data analysis	数据分析
data compression	数据压缩
data link layer	数据链路层
data mining	数据挖掘
data security	数据安全
data structure	数据结构
database application	数据库应用
default margin	默认边距
digital camera	数码相机
digital model	数字模型
digital signal	数字信号
discrete frame	离散帧
display screen	显示屏
distributed computing	分布式计算
DVD player	DVD 播放器
dynamic HTML	动态超文本标记语言
educational institution	教育机构
electronic circuit	电子电路
electronic commerce	电子商务
electronic mail	电子邮件
external storage device	外存设备
face recognition	人脸识别
fiber optics	光纤
flat surface	平面
flush-left	左对齐
font specifications	字体规范
geographic location	地理位置
government agency	政府机构
graphical object	图形对象
handheld computer	手持计算机
handwriting recognition	手写识别
heap sort	堆排序
hierarchical file structure	层次文件结构
high-level programming language	高级编程语言
high-performance computing	高性能计算
home page	首页

human intervention	人为干预
image dimension	图像尺寸
impact printer	击打式打印机
ink-jet printer	喷墨打印机
input device	输入设备
integrated circuit	集成电路
internal buses	内部总线
internet address	因特网地址
inverted tree	倒置树
is similar to	与……类似
Java applet	Java 小程序
laser printer	激光打印机
left justify	左对齐
linked list	链表
lossless compression	无损压缩
low-level language	低级语言
machine language	机器语言
main memory	主存
make a distinction between…and…	对……和……加以区别
make prediction	做出预测
matching and verification	匹配和验证
mechanical device	机械装置
memory address	内存地址
memory buses	内存总线
mobile device	移动设备
mobile phone	移动电话
modulator-demodulator	调制器和解调器
moving image	动态图像
multimedia kit	多媒体套件
network organization	网络组织
network protocol	网络协议
network topology	网络拓扑结构
neural network	神经网络
object code	目标代码
one-dimensional array	一维数组
output device	输出设备
packet protocol	分组协议

packet switching	分组交换
page length	页长
parallel processing algorithm	并行处理算法
pared-down version	精简版
pattern recognition	模式识别
peripheral device	外围设备
personal computer	个人计算机
pervasive computing	普适计算
physical topology	物理拓扑
pointing device	点击设备
protocol stack	协议栈
real time	实时
search engine	搜索引擎
shopping cart	购物车
short for	是……的简称（简记）
smart card	智能卡
sound card	声卡
source code	源代码
star topology	星型拓扑结构
style sheet	样式表
synonym for	是……的同义词
system parameter	系统参数
systems software	系统软件
telecommunication equipment	电信设备
telephone wire	电话线
thermal printer	热敏打印机
traffic cop	交通警察
transaction processing	交易处理
tree structure	树结构
twisted-pair cable	双绞线电缆
two-dimensional array	二维数组
variable names	变量名
video conferencing	视频会议
virtual memory	虚拟内存
virtual reality	虚拟现实
vise versa	反之亦然
voice recognition	语音识别

Web authoring	Web 创作
Web browser	Web 浏览器
Web documents	Web 文档
Web page template	网页模板
Web server	Web 服务器
word processor	文字处理软件
word wrap	自动换行

缩 写 表

ADC—Analog-to-Digital Converter	模数转换器
AI—Artificial Intelligence	人工智能
ASCII—American Standard Code for Interchange Information	美国信息交换标准码
CD-ROM—Compact Disk-Read-Only Memory	只读碟
CPU—Central Processing Unit	中央处理器
CSS—Cascading Style Sheets	层叠样式表
DAC—Digital-to-Analog Converter	数模转换器
DNS—Domain Name System	域名系统
FIFO—first in, first out	先进先出
FTP—file transfer protocol	文件传输协议
GUI—Graphical User Interface	图形用户界面
HTML—HyperText Markup Language	超文本标记语言
HTTP—HyperText Transfer Protocol	超文本传输协议
IC—Integrated Circuits	集成电路
IDE—integrated development environment	集成开发环境
ISP—Internet service provider	因特网服务提供者
LAN—Local Area Network	局域网
LIFO—Last-In, First-Out	后进先出
MAN—Metropolitan Area Network	城域网
NIC—Network Interface Card	网络接口卡
NLP—Natural Language Processing	自然语言处理
NOS—Network Operating System	网络操作系统
OCR—Optical Character Recognition	光学字符识别
OOP—Object-Oriented Programming	面向对象程序设计
OS—Operating System	操作系统
PDA—Personal Digital Assistant	个人数字助理
RAM—Random Access Memory	随机存储器

ROM—Read-Only Memory	只读存储器
SGML—Standard for General Markup Language	通用标记语言标准
TCP/IP—Transfer Control Protocol/Internet Protocol	TCP/IP 协议
URI—Uniform Resource Identifier	统一资源标识符
URL—Uniform Resource Locator	统一资源定位符
WAN—Wide Area Network	广域网
WWW—World Wide Web	万维网
XML—Extensible Markup Language	可扩展标记语言

图书资源支持

感谢您一直以来对清华版图书的支持和爱护。为了配合本书的使用,本书提供配套的资源,有需求的读者请扫描下方的"书圈"微信公众号二维码,在图书专区下载,也可以拨打电话或发送电子邮件咨询。

如果您在使用本书的过程中遇到了什么问题,或者有相关图书出版计划,也请您发邮件告诉我们,以便我们更好地为您服务。

我们的联系方式:

地　　址:北京海淀区双清路学研大厦 A 座 707

邮　　编:100084

电　　话:010-62770175-4604

资源下载:http://www.tup.com.cn

电子邮件:weijj@tup.tsinghua.edu.cn

QQ:883604(请写明您的单位和姓名)

用微信扫一扫右边的二维码,即可关注清华大学出版社公众号"书圈"。

资源下载、样书申请

书圈